UNDERSTANDING CROSS-CULTURAL NEUROPSYCHOLOGY

Understanding Cross-Cultural Neuropsychology thoroughly examines the meaning of culture in the context of neuropsychology, focusing on the fundamental neuroscience underlying how different aspects of culture influence neuropsychological test performance, and how that is related to brain function. It explores in detail the relationship between brain activity and culture, and the influence of various cultural, educational, and linguistic factors on neuropsychological test performances across various cognitive domains.

Written by leading researchers in cross-cultural neuropsychology, the book first introduces the basic concepts in the field. It goes on to focus on the influence of cultural variables on specific domains of cognition, including perception, attention, memory, language, and executive functions. It also explores the implications of cross-cultural neuropsychology in practice, including a focus on test adaptation, the use of interpreters, the influence of acculturation, and the practice of neuropsychological rehabilitation in different cultural settings.

This book is essential reading for neuropsychologists and related practitioners working with culturally diverse clients, who need a good grasp of the cultural impacts on neuropsychological test performance when assessing clients from different cultural, linguistic, and educational backgrounds. It is also valuable for neuropsychologists in countries around the world who need a means of understanding the ways in which their culture impacts the performances of their clients on tests, which have been mostly developed in the U.S. or other Western cultures.

Alberto Luis Fernández is the Head of the Research Department at the Catholic University of Córdoba in Argentina, where he also teaches neuropsychology and psychometrics. He has worked in cross-cultural neuropsychological test adaptation and development and published several articles on the topic and is currently developing a cross-cultural test (The Multicultural Neuropsychological Scale) whose preliminary results have been published. He has been in collaboration with Dr. Jonathan Evans for the last three years in different projects on the topic, among them the organization of a workshop in Chile and the presentation of a course in the INS meeting in Washington in 2018.

Jonathan Evans moved in 1991 to work at the MRC Applied Psychology Unit (now MRC Cognition and Brain Sciences Unit) in Cambridge. In 1996 he became the founding Clinical Director of the Oliver Zangwill Centre for Neuropsychological Rehabilitation in Ely. In October 2003 he moved to his current position at the University of Glasgow, where he is Professor of clinical Neuropsychology and Program Director for the MSc in Clinical Neuropsychology.

Current Issues in Neuropsychology
Series Editor: Jon Evans
University of Glasgow, Glasgow, U.K.

Current Issues in Neuropsychology is a series of edited books that reflect the state-of-the-art in areas of current and emerging interest in the psychological study of brain damage, behavior, and cognition.

Each volume is tightly focused on a particular topic, with chapters contributed by international experts. The editors of individual volumes are leading figures in their areas and provide an introductory overview of the field.

Each book will reflect an issue, area of uncertainty or controversy, with contributors providing a range of views on the central topic. Examples include the question of whether technology can enhance, support or replace impaired cognition, and how best to understand, assess and manage alcohol-related brain damage, and how to maximise new learning in people with memory impairment.

Published titles in the series

Assistive Technology for Cognition
Edited by Brian O'Neill and Alex Gillespie

Alcohol and the Adult Brain
Edited by Jenny Svanberg, Adrienne Withall, Brian Draper, and Stephen Bowden

Errorless Learning in Neuropsychological Rehabilitation
Edited by Catherine Haslam and Roy P.C. Kessels

Psychological Therapies in Acquired Brain Injury
Edited by Giles N. Yeates and Fiona Ashworth

Understanding Cross-Cultural Neuropsychology: Science, Testing, and Challenges
Edited by Alberto Luis Fernández and Jonathan Evans

UNDERSTANDING CROSS-CULTURAL NEUROPSYCHOLOGY

Science, Testing, and Challenges

Edited by Alberto Luis Fernández and Jonathan Evans

LONDON AND NEW YORK

Cover image: Getty Images

First published 2022
by Routledge
2 Park Square, Milton Park, Abingdon, Oxon OX14 4RN

and by Routledge
605 Third Avenue, New York, NY 10158

Routledge is an imprint of the Taylor & Francis Group, an informa business

British Library Cataloguing-in-Publication Data
A catalogue record for this book is available from the British Library

Library of Congress Cataloging-in-Publication Data
A catalog record has been requested for this book

ISBN: 978-0-367-50840-1 (hbk)
ISBN: 978-0-367-50838-8 (pbk)
ISBN: 978-1-003-05149-7 (ebk)

DOI: 10.4324/9781003051497

Typeset in Bembo
by MPS Limited, Dehradun

CONTENTS

CONTRIBUTORS

Gerald H. Burgess, Salomons Institute for Applied Psychology, Canterbury Christ Church University, U.K.

Sumita Chatterjee, Institute of Health & Wellbeing, University of Glasgow, U.K.

Lori Della Malva, Psychology Department, The Ottawa Hospital, Canada

Aparna Dutt, Neuropsychology & Clinical Psychology Unit, Duttanagar Mental Health Centre, Kolkata 700 077, India

Jonathan Evans, Institute of Health & Wellbeing, University of Glasgow, U.K.

Kara Eversole, Department of Graduate Psychology, James Madison University, U.S.A.

Alberto Luis Fernández, Neuropsychology Department, Universidad Católica de Córdoba; Psychometrics Department, Universidad Nacional de Córdoba, Argentina

Daryl Fujii, Veterans Affairs Pacific Island Health Care Services, U.S.A.

Chih-Mao Huang, Department of Biological Science and Technology, National Yang Ming Chiao Tung University, Taiwan

Hsu-Wen Huang, Department of Linguistics and Translation, City University of Hong Kong, Hong Kong

Hajin Lee, School of Rehabilitation, Faculty of Medicine, University of Montreal, Montreal, Quebec, Canada

Karen K. Leung, Department of Family Medicine, University of Alberta, Edmonton, Alberta, Canada.

Adriana Macías Strutt, Department of Neurology, Section of Neuropsychology, Baylor College of Medicine, Houston 77030, U.S.A.; Department of Psychiatry & Behavioral Sciences, Baylor College of Medicine, Houston 77030, U.S.A.

Beatriz MacDonald, Department of Pediatrics, Section of Psychology, Baylor College of Medicine, Houston 77030, U.S.A.

Bernice A. Marcopulos, Department of Graduate Psychology, James Madison University, Department of Psychiatry and Neurobehavioral Sciences, University of Virginia School of Medicine, U.S.A.

Takahiko Masuda, Psychology Department, University of Alberta, Edmonton, Alberta, Canada

Matthew J. Russell, School of Public Policy, University of Calgary, Calgary, Alberta, Canada

Octavio A. Santos, Department of Psychology, The Ottawa Hospital, Canada

Yi Wen Tan, Lee Kuan Yew School of Public Policy, National University of Singapore, Singapore

Jill Winegardner, Director of Neuropsychological Rehabilitation, Neurological Institute, University Hospitals Cleveland Medical Center, Cleveland, OH, U.S.A.

INTRODUCTION

Cross-cultural neuropsychology is an emerging specialty within the field of neuropsychology. Although the issues addressed in this field have been discussed for over four decades, recently the field has commanded greater attention. A quick indicator of current interest in the field is reflected in the results of a Google Scholar search using the term "cross-cultural neuropsychology", which produces approximately 80,000 results. If the search is narrowed to the period between 1981 and 2021 there are approximately 32,800 results. But during the period 2018–2021, there are approximately 17,400 results, indicating that 53% of all outputs from the last 40 years were produced in the last four years. The appearance of journals such as "Culture & Brain" and several books on the topic are also proof of the relevance of the topic. Moreover, the "Cross-cultural Neuropsychology" Special Interest Group (SIG) of the International Neuropsychological Society (INS) was one of the first SIG's to be established.

Although several books on the topic have been published in the last years, most have been dedicated to the acknowledgment of cultural differences and the practice of neuropsychology with clients from different cultures. However, there are very few chapters that examine the hard science underlying these cultural differences, i.e., the relationship between brain activity and culture and on the influence of "cultural" variables on test performance. To better serve our global community we need to improve our understanding of how different aspects of culture influence neuropsychological test performance and how that is related to brain functioning. This book covers a range of topics, from a description of the current challenges in the field, to the most up-to-date information on the findings of the specific influence of different cultural variables on the neuropsychological performance. The first three chapters describe basic concepts in the field including a definition of culture and state-of-the-art knowledge of cultural influences on brain functioning. The second section focuses on the influence of cultural variables on specific domains of cognition, including perception, attention, memory, language, and executive functions. In the third section, there is a discussion of some of the implications for everyday practice in neuropsychology, including a focus on test adaptation, the use of interpreters, the influence of acculturation and the practice of neuropsychological rehabilitation in different cultural settings. A final chapter contains a description of possible solutions to current challenges and future directions for the field.

We hope that this book reflects contemporary evidence and thinking on cross-cultural neuropsychology for both researchers and clinicians. This area of study has experienced significant growth in the last few decades, and we expect it to continue growing in the coming years. We hope that this book will inspire new research and methods that might address the numerous unsolved challenges in the field.

Alberto Luis Fernández and Jonathan Evans

PART I

Basic concepts in cross-cultural neuropsychology

1

CHALLENGES FOR NEUROPSYCHOLOGY IN THE GLOBAL CONTEXT

Aparna Dutt, Jonathan Evans,
and Alberto Luis Fernández

Introduction

An unprecedented acceleration in the movement of people within and across national borders has led to a massive increase in cultural diversity within countries across the globe. The world's population is also aging, increasing the burden of neurological conditions such as dementia and stroke. Consequently, there is an increasing need for culturally appropriate and equitable neuropsychology services right across the world. This is a challenge for countries without an established tradition of neuropsychology research and practice who need to develop services for the majority culture, as well as for minority groups. It is also a challenge for countries with well-established neuropsychology services who need to ensure equity of service to all members of the community. Sociocultural and socio-economic inequalities within and between nations contribute to a higher pre-valence of conditions such as hypertension, diabetes, and HIV/AIDS (Corsi & Subramanian, 2019; Hajizadeh et al., 2014; Stronks et al., 2013) that affect cog-nitive functioning, resulting in a need for neuropsychological services (Rivera Mindt et al., 2010). However, the vast majority of the world's population lack access to neuropsychological services for a myriad of reasons (Allott & Lloyd, 2009; Joosub, 2019; Romero et al., 2009).

Providing culturally competent neuropsychological assessment is seriously challenging and there are many factors that contribute to inequalities in neu-ropsychological service provision. The challenges in meeting the needs for neu-ropsychological rehabilitation are no less prevalent across the globe (See Chapter 13).

The COVID-19 pandemic has further highlighted these challenges. There is increasing evidence of long-term neurological and neuropsychological sequelae due to severe COVID-19 infection (Kirby, 2020). Black and minority ethnic

DOI: 10.4324/9781003051497-2

populations in western countries are disproportionately affected by COVID-19, which has also been devastating in many Low and Middle-Income Countries (LMICs). The pandemic highlights the growing and urgent need to understand, identify and effectively address the challenges in delivering culturally appropriate neuropsychological assessment and neuropsychological rehabilitation.

In this chapter, we will discuss the challenges neuropsychology is facing in the delivery of effective services to everyone in need of them globally, with a focus on neuropsychological assessment. In Chapter 14, we will discuss some practical ways to respond to the challenges.

The challenges

A complex set of interlinked and interdependent challenges involving individuals, health care systems, service providers, neuropsychological training, and neuropsychological tests contributes to inequalities in neuropsychological assessment quality and quantity. These issues affect migrants (immigrants, refugees, and asylum seekers), ethnic minorities, and indigenous groups in regions with neuropsychological services that are well-established for the ethnic majority (e.g., North America, Western Europe, Australia, and New Zealand). In countries where neuropsychology is less established, particularly LMICs, these challenges are relevant to both the ethnic majority and minority groups. A model illustrating these complex interacting factors is available at (https://drive.google.com/file/d/1HcqldEVOxcaimumk1gIAWY_KEo3Hz9nl/view?usp=sharing).

Challenges in access to neuropsychological services – the five "A's"

We suggest that the main challenges that create inequalities in access to neuropsychological services to the aforementioned groups relate to: (a) awareness, (b) acceptability, (c) availability, (d) accessibility, and (e) affordability. The interplay of these challenges leads to under-diagnosis and under-treatment of neuropsychological disorders.

Awareness and acceptability

Lack of awareness of, and stigma associated with, neurological conditions like dementia (Aghvinian et al., 2020; Kenning et al., 2017; Nielsen & Waldemar, 2016), and lack of awareness of potential benefits of neuropsychological services act as barriers to access. These barriers are particularly significant in LMICs, and to many immigrants and ethnic minority groups in countries where neuropsychology is established (Romero et al., 2009), and are likely to be evident in other marginalized populations as well. Even when migrants and people from ethnic minorities are aware of neuropsychological services, personal and environmental factors, and issues affecting the relationship between the patient and the healthcare provider can lead to poor acceptability and reduced help-seeking of neuropsychological services (Leong & Kalibatseva, 2011; Terrell & Terrell, 1983). In contrast, in many non-western countries, particularly in Asia, the acceptability of neuropsychological services is poor due to the dominance of the medical model and a lack of integration of neuropsychological services in the healthcare system (Fujii, 2011). Together, lack of awareness and poor acceptability leads to reduced help-seeking and a low rate of referral for neuropsychological assessment and rehabilitation.

Availability

Perhaps the biggest challenge is the paucity of skilled neuropsychologists across the globe (Grote & Novitski, 2016). The availability of neuropsychologists with relevant knowledge and clinical skills training is often disproportionate to the population (Kasten et al., 2021). These disparities are most evident in LMICs, especially in Asia and Africa, where no specialist training or accredited qualification exists (Ponsford, 2017) but also evident in some European (Hokkanen et al., 2020), and other High-Income Countries (HICs). There is also a severe shortage of qualified culturally and linguistically diverse neuropsychologists to work with international migrants and ethnic minorities in countries where neuropsychology is well-established, which may impede access to a fair neuropsychological assessment in this group (Rivera Mindt et al., 2010). The situation is similar in low-resourced multicultural and multilingual settings like South Africa (Laher & Cockcroft, 2017) and India.

High-level academic training programs in clinical neuropsychology are available in many HIC's (Hokkanen et al., 2020). However, lack of in-depth theoretical and supervised clinical skills training with diverse cultural groups in the training programs is another major barrier to accurate neuropsychological assessment in ethnic minority patients (Baber, 2020; Elbulok-Charcape et al., 2014; Franzen et al., 2020a).

Accessibility

Reliable neuropsychological services, even when available, may not be accessible to all segments of the population. Due to inequitable access to specialists (neurologists and psychiatrists) (Saadi et al, 2017), many immigrants and people from ethnic minorities and other socially disadvantaged groups are less likely to be referred to neuropsychologists. Access to services is also urban biased, limiting access to those from rural communities.

Affordability

Lastly, apart from a few countries where neuropsychological services are covered by health services or by health insurance, unaffordability also limits the utilization of neuropsychological services, particularly when there are differences in service provision between public and private hospital sectors in some countries like South Africa (Watts & Shuttleworth-Edwards, 2016) and India. Additionally, ethnicity-related disparities in health insurance in HICs like the U.S.A. (Rivera Mindt et al., 2010) also restrict access to neuropsychological services.

Sociocultural factors that challenge the validity of neuropsychological assessment

From the clinician's side: cultural competence

Multicultural competence is a fundamental skill for neuropsychologists working with patients in cross-cultural situations. Clinical neuropsychology is cross-cultural "when there are significant cultural or language differences between the examiner, examinee, informants, tests, and/or social context" (Judd et al., 2009). Based on this definition, multicultural competence in neuropsychology can broadly be considered as the clinician's (a) awareness, knowledge and attitude regarding the patient's culture and (b) the skills, knowledge and experience necessary to communicate effectively, understand the way the symptoms may manifest, conduct the clinical interview, tailor the neuropsychological assessment (appropriate test selection, adaptation/translation and administration of tests, accurate scoring, interpretation, and case formulation), effectively communicate the neuropsychological findings with the patient, family and relevant others, and execute neuropsychological interventions that meet the social, cultural and linguistic needs of the patient.

Calls to increase cultural competence in neuropsychological practice have received mounting attention over the past decade (Rivera Mindt et al., 2010). To date, however, lack of cultural competence in clinicians due to lack of multicultural training continues to persist leading to inequalities in neuropsychological service provision, assessment and rehabilitation worldwide.

From the patient's side

There is substantial evidence that a person's cultural background and environment influences cognition. Although cognitive functions are considered universal, there may be cultural diversity in their representation. The strategies required to carry out these functions may also vary depending on our cultural experience (Ardila, 2005; Rosselli & Ardila, 2003).

Cross-cultural differences have been observed on a range of neuropsychological measures, including tasks assessing speed of processing, which have been attributed to subjective cultural interpretations relating to the construct of time and speed (Agranovich et al., 2011). Superior performances on tasks assessing working memory (i.e., digit span) in cultures where digits have shorter pronunciation durations (Chan & Elliott, 2011) are unsurprising but have important implications for use of digit span test norms. Differential/varying word production on verbal fluency tasks have also been attributed to linguistic properties (i.e., the shorter versus the longer syllabic length) (Ardila, 2020), or substantial difficulty in non-alphanumeric languages (e.g., Arabic or Kanji).

While the precise contribution of culture to neuropsychological test performance remains to be determined, several cultural and contextual factors may bias assessment outcomes, including familiarity, education/literacy, language, cultural values, socioeconomic status, geographic region, dwelling condition, immigration context and level of acculturation, and differences in cognitive processing styles.

Educational experiences warrant close consideration due to heterogeneity across and within nations, and also across different migrant and ethnic minority groups. Research has established that a higher level of education is associated with better performance on neuropsychological tests (Ardila et al., 2010). Given that a majority of the neuropsychological tests have been standardized within educated Western populations, interpretation of what constitutes a normal test performance within low educated culturally diverse groups is a global problem. Individuals who differ demographically from normative samples of several commonly used tests tend to perform in the impaired range. It is inaccurate to assume that individuals with limited levels of education are somehow less intelligent or less able in relation to underlying cognitive processes. Rather, it may be more accurate to suggest that for this group the acquisition of cognitive skills occurs within the context of practical and procedural modes of learning and such skills may not be readily measurable by current tests. The difference observed in psychometric assessment is also attributed, at least in part, to exposure to school curricula and styles of education that foster cognitive skills like abstract problem solving which are heavily embedded within IQ tests (Nell, 2000). It is important to consider both quantity and quality of education, with quality often being a better predictor of cognitive performance than years of education (Dotson et al., 2009). Therefore, comparisons between educationally dissimilar and/or disadvantaged culturally diverse groups against a Western benchmark, are in most cases, simply not appropriate. However, neuropsychological data regarding the role of quality of education across culturally diverse groups is sparse. Although reading achievement is commonly used to assess quality (Manly et al., 2002), the challenge remains in measuring quality due to several factors. The importance of education in neuropsychological test performance is discussed further in Chapter 4 (Fernandez).

Illiteracy complicates the situation. Although illiteracy is reducing, it still affects a significant proportion of the world's population. Illiteracy may limit accessibility to neuropsychological services and is particularly relevant to neuropsychological assessment of older adults with suspected dementia in LMICs and from ethnic minority groups (see Kosmidis 2018). Due to reduced access of illiterate individuals to neuropsychological services, the clinician may not have a good understanding of the ways illiterate individuals and their families experience and understand cognitive problems, which is relevant to interpreting the reporting of problems by illiterate individuals and their families. Illiteracy may also prevent individuals from engaging fully in the assessment process. There may be a lack of familiarity with the testing situation, materials and procedures, difficulty in understanding the purpose of assessment and limitations in comprehending test instructions (Ardila et al., 2010; Ganguli et al., 1995). Due to unfamiliarity with the testing situation, illiterate individuals, particularly women may refuse to participate in the assessment process without the presence of a family member. To perform optimally on certain cognitive tasks, illiterate people may also apply strategies that enable them to handle everyday requirements. For example, they may use their fingers during serial subtraction tasks or may repeat words after the examiner during word list learning trials. An inexperienced clinician may not be aware of such strategies, and may unnecessarily prevent them from applying the strategies which otherwise would help them to perform optimally. Heterogeneity within the illiterate population in the context of literacy level (pure/functional), language, race/ethnicity, dwelling, socioeconomic status, occupation (high/low skilled illiterate laborers) migration/immigration, refugee status, tribal and indigenous populations and also gender issues only adds to the complexity. Furthermore, elderly illiterate individuals in some LMICs often do not know their year of birth which leads to uncertainty as to which age-based norms to use for test interpretation (Noroozian et al., 2014).

Considering all these limitations it is difficult to know what current neuropsychological tests are measuring in the illiterate population. It might be the case that, in a specific test, the purportedly measured construct is not being measured in this population, or that the constructs measured in the literate population have a different representation in the illiterate population.

Linguistic diversity proves to be equally challenging in two other contexts: (a) when the patient and the clinician do not understand or speak the same language and (b) formulating/implementing neuropsychological assessment in a bilingual/ multilingual context. Challenges may arise due to (a) limitation in language competency of the clinician, (b) linguistic diversity in the same language across different countries (e.g., Spanish, Bengali), (c) limited availability of translated, adapted, and validated versions of the tests (language and other cognitive domains) in the language/languages in which the assessment needs to be conducted and (d) limited knowledge of the nature of linguistic impairments in aphasias in non-European languages which may interfere with appropriate task selection. Involving a professional interpreter fluent in the source and the target language may help to address the challenges associated with the clinician's language competency and test translations. However, there are several constraints when working with interpreters (See Chapter 11 – Fujii) including whether they have experience working in a neuropsychological context. Furthermore, in many non-western multilingual countries like India and South Africa, professional interpreters are not readily available or almost unheard of in healthcare settings (Claassen et al., 2017; Narayan, 2013). Bilingual family members often serve as ad-hoc translators which can create additional issues. However, the question arises whether involving a family member is arguably preferable to conducting no assessment at all in such settings?

Differences in processing style across cultures are another important factor. Our cultural norms, values, and experiences determine the way we process information (Nisbett et al., 2001). Based on around five major frameworks, cultural values may vary across four to nine different dimensions such as individualism-collectivism, power distance, short-term-long-term orientation, uncertainty avoidance, and masculinity-femininity (Hofsted, 1980, as cited in Thomas, 2008). The individualism-collectivism dimension, one of the widely studied cultural values, has been implicated as having possible consequences on psychological processing and provides "a powerful explanatory tool for understanding the variability in the behavior of individuals in different parts of the world" (Oyserman et al., 2002). Based on this dimension, societies have been categorized into either individualist or collectivist. Studies mostly comparing East Asians (collectivist) and Americans (individualist) have explored this cultural dimension and its impact on attentional control (Hedden et al., 2008), facial processing (Goh et al., 2010) and object re-cognition (Gutchess et al., 2006), concluding that Easterners tend to process in-formation more holistically and attend to contextual information more, whereas, Westerners are more analytical and process information independent of context. The individualism-collectivism dimension and its relevance to neuropsychology are discussed further in Chapters 3 (Huang) and 6 (Chatterjee).

Finally, for international migrants and people from ethnic minorities in HICs with well-established neuropsychology services, conditions surrounding the migration process remain a major challenge for neuropsychological assessment. Two important cultural influences deserve attention in this respect. First, the context of migration (Fujii, 2018), which varies considerably across countries, is particularly relevant for first-generation migrants. If this is not carefully considered when interpreting test findings, it may often contribute to a misdiagnosis. For example, stress and trauma, often associated with the migratory process (pre and post), can result in PTSD, especially in refugees and asylum seekers (Bustamante et al., 2018). In refugees and asylum seekers, torture and injuries such as TBI are also common (McPherson, 2019). Due to an overlap of symptoms between some forms of TBI and PTSD, the differential diagnosis between the two conditions is often chal-lenging. Neuropsychological assessment may be particularly useful in differ-entiating the two conditions, but may prove even more challenging in refugees (Veliu & Leathem, 2017), especially if the context of migration is not taken into account. Second, the level of acculturation may potentially impact neu-ropsychological test performance in international migrants and people from ethnic minorities (Tan et al., 2021). Acculturation is a complex and multidimensional process of cultural and psychological change following continuous exposure to and experience with the host culture that results in varying levels of assimilation, se-paration, integration, and marginalization among migrants and ethnic minorities (see Chapter 12). Since educational experience impacts familiarity with test-taking attitudes, formats and skills, level of acculturation which varies with educational experience in turn may lead to varying levels of performance in people with different migration and ethnic backgrounds. Hence, challenges with an accurate diagnosis may arise if the neuropsychological assessment of migrants and people from ethnic minorities is conducted based simply on their cultural background and not on a formal evaluation of the level of acculturation.

In general, higher levels of acculturation to the mainstream culture are associated with better test performance, specifically on tasks of language, attention, processing speed, and executive functions (Al-Jawahiri & Nielsen, 2021; Kennepohl et al., 2004; Razani et al., 2007). More specifically, however, widely-used proxy acculturation factors in the context of neuropsychological assessment have differential effects on test performance. Some of these factors include language proficiency, educational experience (within or outside the host society; duration of education within the host society), age at arrival and proportion of lifetime residency in the host society, cultural knowledge, cultural identity, and competence. It has been suggested that the level of acculturation should be a known quantity, and should also be used as a normative variable (Razani et al., 2007). However, inconsistencies have also been reported (Rivera Mindt et al., 2010) and the mechanisms by which these factors affect neuropsychological performance is also not yet clear.

On the one hand, our knowledge is limited regarding the neuropsychological implications of cultural differences in cognitive processing styles and acculturation. On the other, cultural changes within non-western societies due to rapid globalization, and an exponential rise in the virtual influence of westernization is further compounding the situation. First, cultural shifts toward individualism have become particularly evident in Japan (Ogihara, 2018) as well as India, Bangladesh, Nigeria, Israel, Chile, and Argentina (Inkeles, 1975). Second, acculturation is no longer an immigration-based phenomenon. Remote acculturation (Ferguson & Bornstein, 2012) to the western culture can occur even in those residing in non-western societies. The question arises whether such changes have different neuropsychological implications and will impact clinical practice? Acculturation may make traditional western tests more relevant for a wider range of people, but remote acculturation may mean that measures of length of stay in the host country will no longer be accurate measures of acculturation.

Socio-cultural factors do not work in isolation. The interaction between the socio-cultural factors and demographic factors like age, educational experience, and sex is not straightforward and varies across individuals and cultures. Stroke, dementia, and acquired brain injuries may add a layer of complexity to the interaction. Not addressing the implications of such complex interactions on the tests and norms developed for the western population increases the risk of misdiagnosis, more specifically overestimation of cognitive impairment in culturally, educationally, and linguistically diverse patients. However, there exists a huge gap between "knowing" about the impact of sociocultural issues and what many neuropsychologists actually do during routine neuropsychological practice and research involving test adaptation and norm development for diverse populations.

Assessment of culturally, linguistically, and educationally diverse people with existing neuropsychological tests and norms that challenge the validity of neuropsychological assessment

It is now widely acknowledged that many neuropsychological tests and norms developed for use with Western mono-cultural, educated, and English-speaking groups are inappropriate for use with culturally, linguistically, and educationally diverse groups across the world. This issue has been recognized by the American Academy of Clinical Neuropsychology, who report that by 2050, 60% of the American population will be "un-testable" with current toolkits of largely mono-lingual, mono-cultural neuropsychological assessment strategies. The latter sentiment has been echoed throughout the global neuropsychological community.

Test appropriateness refers not only to the semantic and linguistic content of the tests in a particular culture but also to cognitive constructs. For example, the digit span test despite being widely used, possibly does not measure the construct attention/working memory across all cultures. Despite cultural differences on several neuropsychological tests, only a few studies have examined the extent to which different tests or test batteries measure the same constructs across diverse cultural groups (Helms-Lorenz et al., 2003). Tests might therefore lack clinical utility for culturally diverse groups due to the high probability of false-positive diagnosis (Daugherty et al., 2017). This has been recognized as a significant barrier to reliable clinical practice, if clinicians use culturally inappropriate tests and norms. They may also deviate from standard administration procedure in such contexts which runs the risk of compromising the reliability of the test score. Despite the large corpus of research surrounding the issue of cultural in-appropriateness of tests and norms, comprehensive "gold standard" neu-ropsychological assessment tools tend to be inaccessible and a vast majority of the world's population continues to remain underserviced (Brickman et al., 2006; Ponsford, 2017). The available indigenous and locally developed tools have not yet been able to fill the gap (Chan et al., 2003; Rock & Price, 2019). All this is partly due to the challenges in translating, adapting/developing, and validating tests. A lack of rigor and transparent reporting regarding the translation/adaptation process may also raise concerns regarding the validity of the tests. Additionally, even when a test proves culturally appropriate, not modifying the scoring criteria taking into consideration the cultural influence on test performance may also lead to misdiagnosis. In a recent study that examined the appropriateness of the clock drawing test in a Bengali-speaking population in India, it was recognized that the Rouleau scoring system had to be modified with credit being given for numbers written partly in English and partly in Bengali scripts (Crombie, 2021).

Reliable and valid neuropsychological measures are even fewer for non-western low educated and illiterate populations (Franzen et al., 2020b). Also, since literacy skills differ substantially among countries with similar levels of formal education, and occupation may not require the use of literacy skills, tests which prove culturally appropriate and clinically useful for people who are illiterate and low educated in one population may not be appropriate for another.

Previous attempts to ameliorate the effects of culture and education have resulted in restricting the testing process to non-verbal tasks. However, this notion is now considered to be problematic, with research identifying that culture, and level/quality of education still have a substantial impact on non-verbal tasks, including visuoperceptual, visuospatial and executive functions (for a review, see Rosselli & Ardila, 2003). Performance of healthy elderly illiterate non-western immigrants has been noted to fall within the impaired ranges on tasks assessing drawing reproduction of simple and complex geometrical figures (Nielsen & Jorgensen, 2013). It proves challenging to determine whether a particular performance on drawing tasks (qualitative or quantitative) is due to a lack of task familiarity, differences in learning opportunities, including the use of pens or pencils, or underlying brain damage (Ardila et al., 2010). Cultural bias on non-verbal tests has been observed in the clinical population also. For comparable sociodemographic and clinical characteristics, Canadians with Parkinson's Disease (PD) born in Anglosphere countries outperformed first-generation immigrants in Canada (born outside Anglosphere countries) on the Rey-Osterrieth Complex Figure (ROCF) Copy task, basic visuoperceptual tasks – Judgment of Line Orientation (JLO), silhouettes and object decision tests in the Visual Object and Space Perception Battery (VOSP) – and executive measures – Wisconsin Card Sorting Test (WCST) and Matrix Reasoning (Statucka & Cohn, 2019).

In the case of norms, the use of population-based norms – mostly age and education adjusted – that have been developed for use within English-speaking culturally diverse nations, are likely to be problematic for use with different ethnic groups that are under-represented in such norms. Methodological issues have impacted the production of clinically valid measures during attempts to carry out test adaptation, norming, and/or sampling according to long-standing traditional techniques in culturally diverse populations. The use of "mixed bag" norms has been examined in some detail by Shuttleworth-Edwards (2016) with regards to the production of population-based South African (SA) standardizations of the WAIS-IV, and by Rivera Mindt and colleagues (2010) with regards to demographic corrections for ethnic minorities in the United States for the WAIS-III and WMS-III. Both these papers question the utility of the norms and argue that they have limited relevance for any of the diverse ethnic groups that make up the overall sample as a result of failing to account for several sociocultural factors.

Interestingly, cultural differences have also been observed on tasks that have been marketed by test developers as "culture fair". For example, Messinis and colleagues (2011) showed that if the U.S. normative data were used in Greek nationals (aged between 45–59 with an education level of nine-11 years), a raw score time of 55 seconds for the Color Trails Test (CTT-1) and 122 seconds for CTT-2 would place their performance in the 34th percentile. In contrast, the same raw scores using culturally appropriate normative data would place this performance in the 50th percentile. In other words, the use of U.S. norms would to some extent unfairly penalize Greek nationals and possibly other culturally diverse groups. Also, Al-Jawahiri and Nielsen (2021) found CTT performances to be influenced by the level of acculturation in immigrants from diverse cultural and linguistic backgrounds in Denmark, with faster performances in those with a higher level of acculturation toward the Danish mainstream culture. These findings indicate the need for culturally appropriate norms due to varying levels and different quality/styles of education, varying familiarity with timed procedures, and the subjective cultural importance of completing timed tasks in a speeded manner (Agranovich et al., 2011). Recent global efforts to address this issue have resulted primarily in the production of cross-cultural clinically viable screen measures (Fernandez & Abe, 2018), such as the Common Objects Memory Test (COMT) (Kempler et al., 2010). However, domain-specific cross-cultural measures and norms remain limited.

Conclusion

It is clear that despite some provision and progress in access and reliability of neuropsychological services for culturally, linguistically, and educationally diverse people across the globe, neuropsychology continues to be a "white privilege" (Cory, 2021). The challenges facing the global neuropsychological community concerning training, practice, and research are undeniable. With increasing changes and disparities within cultures due to globalization and westernization, the challenge is becoming even more complex. However, the issue has received international recognition leading to a call for a neuropsychological cultural revolution. Indeed, a lot needs to be done and collective action both at the national and international level will provide the opportunity to promote equality and precision in neuropsychological assessment amid the many diversities. In Chapter 14, we will discuss some practical short- and long-term solutions.

References

Aghvinian, M., Santoro, A.F., Gouse, H., Joska, J.A., Linda, T., Thomas, K.J.F., & Robbins, R.N. (2020). Taking the test: A qualitative analysis of cultural and contextual factors impacting neuropsychological assessment of Xhosa-speaking South Africans. *Archives of Clinical Neuropsychology*, 10.1093/arclin/acaa115.

Agranovich, A.V., Panter, A.T., Puente, A.E., & Touradji, P. (2011). The culture of time in neuropsychological assessment exploring the effects of culture-specific time attitudes on timed test performance in Russian and American Samples. *Journal of the International Neuropsychological Society*, 17, 692–701. 10.1017/S1355617711000592.

Al-Jawahiri, F., & Nielsen, T.R. (2021). Effects of acculturation on the Cross-cultural Neuropsychological Test Battery (CNTB) in a culturally and linguistically diverse population in Denmark, 14. doi: 10.1093/arclin/acz083.

Allott, K., & Lloyd, S. (2009). The provision of neuropsychological services in Rural/ Regional settings: Professional and ethical issues. *Applied Neuropsychology*, *16*(3), 193–206. 10.1080/09084280903098760.

Ardila, A. (2005). Cultural values underlying psychometric cognitive testing. *Neuropsychology Review*, *15*, 185–195. 10.1007/s11065-005-9180-y.

Ardila, A. (2020). A cross-linguistic comparison of category verbal fluency test (ANIMALS): A systematic review. *Archives of Clinical Neuropsychology*, *35* 205–217. 10.1093/arclin/acz060.

Ardila, A., Bertolucci, P.H., Braga, L.W., Castro-Caldas, A., Judd, T., Kosmidis, M.H., Matute, E., Nitrini, R., Ostrosky-Solis, F., & Rosselli, M. (2010). Illiteracy: The neuropsychology of cognition without reading. *Archives of Clinical Neuropsychology*, *25*, 689–712. 10.1093/arclin/acq079.

Baber, Z. (2020). *An exploration of the perspectives of neuropsychologists working with clients from ethnically, culturally and linguistically diverse backgrounds* (Doctoral dissertation, University of East London). Retrieved from https://repository.uel.ac.uk/item/88861.

Brickman, A.M., Cabo, R., & Manly, J.J. (2006). Ethical issues in cross-cultural neuropsychology. *Applied Neuropsychology*, *13*(2):91–100. 10.1207/s15324826an1302_4.

Bustamante, L.H.U., Cerqueira, R.O., Leclerc, E., & Brietzke, E. (2018). Stress, trauma, and posttraumatic stress disorder in migrants: A comprehensive review. *Revista Brasileira De Psiquiatria*, *40*(2), 220–225. 10.1590/1516-4446-2017-2290.

Chan, A.S., Shum, D., & Cheung, R.W.Y. (2003). Recent development of cognitive and neuropsychological assessment in Asian countries. *Psychological Assessment*, *15*(3), 257–267. 10.1037/1040-3590.15.3.257.

Chan, M.E., & Elliott, J.M. (2011). Cross-Linguistic differences in digit memory span. *Australian Psychologist*, *46*(1), 25–30. 10.1111/j.1742-9544.2010.00007.x.

Claassen, J., Jama, Z., Manga, N., Lewis, M., & Hellenberg, D. (2017). Building freeways: Piloting communication skills in additional languages to health service personnel in Cape town, South Africa. *BMC Health Services Research*, *17*(1), 390. 10.1186/s12913-017-2313-1.

Corsi, D.J., & Subramanian, S.V. (2019). Socioeconomic gradients and distribution of diabetes, hypertension, and obesity in India. *JAMA Network Open*, *2*(4), e190411–e190411. 10.1001/jamanetworkopen.2019.0411.

Cory, J.M. (2021). White privilege in neuropsychology: An 'invisible knapsack' in need of unpacking? *Clinical Neuropsychologist*, *35*(2), 206–218. 10.1080/13854046.2020.1801845.

Crombie., M. (2021) *Examination of the impact of education on cognitive screening tests.* (D Clin Psy thesis, University of Glasgow). Retrieved from https://theses.gla.ac.uk/82193/.

Daugherty, J.C., Puente, A.E., Fasfous, A.F., Hidalgo- Ruzzante, N., & Pérez-Garcia, M. (2017). Diagnostic mistakes of culturally diverse individuals when using North American neuropsychological tests, *Applied Neuropsychology: Adult*, *24*(1), 16–22. 10.1080/23279095.2015.1036992.

Dotson V.M., Kitner-Triolo M.H., Evans M.K., & Zonderman A.B. (2009). Effects of race and socioeconomic status on the relative influence of education and literacy on cognitive functioning. *Journal of the International Neuropsychological Society*, *15*, 580–589. doi: 10.1017/S1355617709090821.

Elbulok-Charcape, M.M., Rabin, L.A., Spadaccini, A.T., & Barr, W.B. (2014). Trends in the neuropsychological assessment of ethnic/racial minorities: A survey of clinical neuropsychologists in the United States and Canada. *Cultural Diversity and Ethnic Minority Psychology*, *20*(3), 353–361. 10.1037/a0035023.

Ferguson, G.M., & Bornstein, M.H. (2012). Remote acculturation: The 'Americanization' of Jamaican islanders. *International Journal of Behavioral Development*, *36*(3), 167–177. doi: 10.1177/0165025412437066.

Fernandez A.L., & Abe J. (2018) Bias in cross-cultural neuropsychological testing: Problems and possible solutions. *Cult. Brain.* doi: 10.1007/s40167-017-0050-2.

Franzen, S., Papma, J.M., van den Berg, E., & Nielsen, T.R. (2020a) Cross-cultural neuropsychological assessment in the European union: A Delphi expert study. *Archives of Clinical Neuropsychology*, 10.1093/arclin/acaa083.

Franzen, S., van den Berg, E., Goudsmit, M., Jurgens, C.K., van de Wiel, L., Kalkisim, Y., Uysal-Bozkir, Ö., Ayhan, Y., Nielsen, T.R., & Papma, J.M. (2020b). A systematic review of neuropsychological tests for the assessment of dementia in non-western, low-educated or illiterate populations. *Journal of the International Neuropsychological Society*, *26*(3), 331–351. 10.1017/S1355617719000894.

Fujii, D. (2011). *The Neuropsychology of Asian-Americans*. Psychology Press.

Fujii, D.E.M. (2018) Developing a cultural context for conducting a neuropsychological evaluation with a culturally diverse client: the ECLECTIC framework, *The Clinical Neuropsychologist*, *32*(8), 1356–1392, DOI: 10.1080/13854046.2018.1435826.

Ganguli, M., Ratcliff, G., Chandra, V., Sharma, S., Gilby, J., Pandav, R., Belle, S., Ryan, C., Baker, C., Seaberg, E., & Dekosky, S. (1995). A Hindi version of the MMSE: The development of a cognitive screening instrument for a largely illiterate rural elderly population in India. *International Journal of Geriatric Psychiatry*, *10*(5), 367–377. 10.1002/gps.930100505.

Goh, J.O.S., Leshikar, E.D., Sutton, B.P., Tan, J.C., Sim, S.K.Y., Hebrank, A.C., & Park, D.C. (2010). Culture differences in neural processing of faces and houses in the ventral visual cortex. *Social Cognitive and Affective Neuroscience*, *5*(2-3), 227–235. 10.1093/scan/nsq060.

Grote, G.L., & Novitski, J.I. (2016). International perspectives on education, training, and practice in clinical neuropsychology: comparison across 14 countries around the world. *The Clinical Neuropsychologist*. 10.1080/13854046.2016.1235727.

Gutchess, A.H., Welsh, R.C., Boduroğlu, A., & Park, D.C. (2006). Cultural differences in neural function associated with object processing. *Cognitive, Affective, & Behavioral Neuroscience*, *6*(2), 102–109. 10.3758/CABN.6.2.102.

Hajizadeh, M., Sia, D., Heymann, S.J., & Nandi, A. (2014). Socioeconomic inequalities in HIV/AIDS prevalence in sub-saharan african countries: Evidence from the demographic health surveys. *International Journal for Equity in Health*, *13*(1), 18. 10.1186/1475-9276-13-18.

Hedden, T., Ketay, S., Aron, A., Markus, H.R., & John D.E. Gabrieli. (2008). Cultural influences on neural substrates of attentional control. *Psychological Science*, *19*(1), 12–17. 10.1111/j.1467-9280.2008.02038.x.

Helms-Lorenz, M., Van de Vijver, F.J.R., & Poortinga, Y.H. (2003). Cross-cultural differences in cognitive performance and Spearman's hypothesis: g or c? *Intelligence*, *31*, 9–29. doi:10.1016/s0160-2896(02)00111-3.

Hokkanen L., Barbosa F., Ponchel A., Constantinou M., Kosmidis M.H., Varako N., Kasten E., Mondini S., Lettner S., Baker G., Persson B.A., Hessen E. (2020) Clinical Neuropsychology as a Specialist Profession in European Health Care: Developing a Benchmark for Training Standards and Competencies Using the Europsy Model? *Frontiers in Psychology*. *11*:559134. doi: 10.3389/fpsyg.2020.559134.

Inkeles A. (1975). Becoming modern: Individual change in six developing countries. *Ethos (Berkeley, Calif.)*, *3*(2), 323–342. 10.1525/eth.1975.3.2.02a00160.

Joosub, N. (2019). How local context influences access to neuropsychological rehabilitation after acquired brain injury in South Africa. *BMJ Global Health*, *4*(Suppl 10), e001353–e001353. 10.1136/bmjgh-2018-001353.

Judd, T., Capetillo, D., Carrión-Baralt, J., Mármol, L.M., Miguel-Montes, L.S., Navarrete, M.G., Puente, A.E., Romero, H.R., Valdés, J., NAN Policy and Planning Committee, & and the NAN Policy and Planning Committee. (2009). Professional considerations for improving the neuropsychological evaluation of Hispanics: A National Academy of Neuropsychology education paper. *Archives of Clinical Neuropsychology*, *24*(2), 127- 135. 10.1093/arclin/acp016.

Kasten, E., Barbosa, F., Kosmidis, M.H., Persson, B.A., Constantinou, M., Baker, G.A., Lettner, S., Hokkanen, L., Ponchel, A., Mondini, S., Jonsdottir, M.K., Varako, N., Nikolai, T., Pranckeviciene, A., Harper, L., & Hessen, E. (2021). European clinical neuropsychology: Role in healthcare and access to neuropsychological services. *Healthcare (Basel), 9*(6), 734. 10.3390/healthcare9060734.

Kempler, D., Teng, E.L., Taussig, M., & Dick, M.B. (2010). The common objects memory test (COMT): A simple test with cross-cultural applicability. *Journal of the International Neuropsychological Society, 16*(3), 537–545. 10.1017/S1355617710000160.

Kennepohl, S., Shore, D., Nabors, N., & Hanks, R. (2004). African American acculturation and neuropsychological test performance following traumatic brain injury. *Journal of the International Neuropsychological Society, 10*(4), 566–577. 10.1017/S135561 7704104128.

Kenning, C., Daker-White, G., Blakemore, A., Panagioti, M., & Waheed, W. (2017). Barriers and facilitators in accessing dementia care by ethnic minority groups: A meta-synthesis of qualitative studies. *BMC Psychiatry, 17*(1), 316. 10.1186/s12888-017-1474-0.

Kirby, T. (2020). Evidence mounts on the disproportionate effect of COVID-19 on ethnic minorities. *The Lancet Respiratory Medicine, 8*(6), 547. 10.1016/S2213-2600(20)30228-9.

Kosmidis, M.H. (2018). Challenges in the neuropsychological assessment of illiterate older adults. *Language, Cognition and Neuroscience, 33*(3), 373–386. 10.1080/23273798.2017.13 79605.

Laher, S., & Cockcroft, K. (2017). Moving from culturally biased to culturally responsive assessment practices in low-resource, multicultural settings. *Professional Psychology, Research and Practice, 48*(2), 115–121. 10.1037/pro0000102.

Leong, F.T.L., & Kalibatseva, Z. (2011). Cross-cultural barriers to mental health services in the United States. *Cerebrum* (New York, NY), *2011*, 5.

Manly, J.J., Jacobs, D.M., Touradji, P., Small, S.A., & Stern, Y. (2002). Reading level attenuates differences in neuropsychological test performance between African American and White elders. *Journal of the International Neuropsychological Society, 8,* 341–348. 10.1017/S1355617702813157.

McPherson, J.I. (2019). Traumatic brain injury among refugees and asylum seekers. *Disability and Rehabilitation, 41*(10), 1238–1242. 10.1080/09638288.2017.1422038.

Messinis, L., Malegiannaki, A.C., Christodoulou, T., Panagiotopoulos, V., & Papathanasopoulos, P. (2011). Colour Trails Test: Normative Data and Criterion Validity for the Greek Adult Population. *Archives of Clinical Neuropsychology, 26,* 322–330. 10.1093/arclin/acr027.

Narayan, L. (2013). Addressing language barriers to healthcare in India. *The National Medical Journal of India, 26*(4), 236.

Nell, V. (2000). *Cross-cultural neuropsychological assessment: Theory and practice.* Mahwah, NJ: Lawrence Erlbaum Associates, Inc.

Nielsen, T.R., & Waldemar, G. (2016). Knowledge and perceptions of dementia and alzheimer's disease in four ethnic groups in Copenhagen, Denmark. *International Journal of Geriatric Psychiatry, 31*(3), 222–230. 10.1002/gps.4314.

Nielsen, T.R., & Jorgensen, K. (2013). Visuoconstructional abilities in cognitively healthy illiterate Turkish Immigrants: A Quantitative and Qualitative Investigation. *The Clinical Neuropsychologist, 27,* 4. 10.1080/13854046.2013.767379.

Nisbett R.E., Peng K., Choi I., Norenzayan A. (2001). Culture and systems of thought: Holistic versus analytic cognition. *Psychological Review, 108*(2), 291–310. 10.1037//0033-295X.108.2.291.

Noroozian, M., Shakiba, A., & Iran-Nejad, S. (2014). The impact of illiteracy on the assessment of cognition and dementia: A critical issue in the developing countries. *International Psychogeriatrics, 26*(12), 2051–2060. 10.1017/S1041610214001707.

Ogihara, Y. (2018). The rise in individualism in japan: Temporal changes in family structure, 1947-2015. *Journal of Cross-Cultural Psychology, 49*(8), 1219–1226. 10.1177/0022022118781504.

Oyserman, D., Coon, H.M., & Kemmelmeier, M. (2002). Rethinking individualism and collectivism: Evaluation of theoretical assumptions and meta-analyses. *Psychological Bulletin, 128*(1), 3–72. 10.1037/0033-2909.128.1.3.

Ponsford, J. (2017). International growth of Neuropsychology. *Neuropsychology, 31*(8), 921–933. 10.1037/neu0000415.

Razani, J., Burciaga, J., Madore, M., & Wong, J. (2007). Effects of acculturation on tests of attention and information processing in an ethnically diverse group. *Archives of Clinical Neuropsychology, 22*(3), 333–341. 10.1016/j.acn.2007.01.008.

Rivera Mindt, R.M., Byrd, D., Saez, P., & Manly, J.J. (2010). Increasing culturally competent neuropsychological services for ethnic minority populations: a call to action. *Clinical Neuropsychology, 24*(3): 429–453. 10.1080/13854040903058960.

Rock, D., & Price, I.R. (2019). Identifying culturally acceptable cognitive tests for use in remote northern Australia. *BMC Psychology, 7*(1), 62. 10.1186/s40359-019-0335-7.

Romero, H.R., Lageman, S.K., Kamath, V., Irani, F., Sim, A., Suarez, P., Manly, J.J., Attix, D.K., participants, T.S., & Summit participants. (2009). Challenges in the neuropsychological assessment of ethnic minorities: Summit proceedings. *Clinical Neuropsychologist, 23*(5), 761–779. 10.1080/13854040902881958.

Rosselli, M. & Ardila, A. (2003). The impact of culture and education on non-verbal neuropsychological measurements: a critical review. *Brain and Cognition, 52*, 326–333. 10.1016/S0278-2626(03)00170-2.

Saadi, A., Himmelstein, D.U., Woolhandler, S., & Mejia, N.I. (2017). Racial disparities in neurologic health care access and utilization in the united states. *Neurology, 88*(24), 2268–2275. 10.1212/WNL.0000000000004025.

Shuttleworth-Edwards, A.B. (2016). Generally representative is representative of none: commentary on the pitfalls of IQ test standardization in multicultural settings. *The Clinical Neuropsychologist, 30*(7), 975–998. 10.1080/13854046.2016.1204011.

Statucka, M., & Cohn, M. (2019). Origins matter: Culture impacts cognitive testing in Parkinson's disease. *Frontiers in Human Neuroscience, 13*, 269. 10.3389/fnhum.2019.00269.

Stronks, K., Snijder, M.B., Peters, R.J.G., Prins, M., Schene, A.H., & Zwinderman, A.H. (2013). Unravelling the impact of ethnicity on health in Europe: The HELIUS study. *BMC Public Health, 13*(1), 402. 10.1186/1471-2458-13-402.

Tan, Y.W., Burgess, G.H., & Green, R.J. (2021). The effects of acculturation on neuropsychological test performance: A systematic literature review. *Clinical Neuropsychologist, 35*(3), 541–571. 10.1080/13854046.2020.1714740.

Thomas, D.C. (2008). Comparing cultures: Systematically describing cultural differences. In *Cross cultural management: Essential concepts* (pp. 47–69). Thousand Oaks, CA: Sage Publications.

Terrell, F., & Terrell, S. (1983). The relationship between race of examiner, cultural mistrust, and the intelligence test performance of black children. *Psychology in the Schools, 20,* 367–369. 10.1002/1520-6807(198307)20:3%3C367::AID-PITS2310200318%3E3.0.CO;2-Y.

Veliu, B., & Leathem, J. (2017). Neuropsychological assessment of refugees: Methodological and crosscultural barriers. *Applied Neuropsychology. Adult, 24*(6), 481–492. 10.1080/232 79095.2016.1201483.

Watts, A.D., & Shuttleworth-Edwards, A.B. (2016). Neuropsychology in South Africa: Confronting the challenges of specialist practice in a culturally diverse developing country. *Clinical Neuropsychologist, 30*(8), 1305–1324. 10.1080/13854046.2016.1212098.

2

OPERATIONALIZING THE CONCEPT OF CULTURE

Alberto Luis Fernández

What is culture?

Culture is a very broad concept. It has been defined in many different ways along several dimensions. Kroeber and Kluckhohn (1967) suggested that there are six major classes of definitions of culture including descriptive, historical, normative, psychological, structural, and genetic. Reviewing all of these definitions is probably a futile exercise for the goals of this chapter but as an example of the diversity of definitions two might be quoted here: "Culture means the whole complex of traditional behavior which has been developed by the human race and is successively learned by each generation. *A culture* is less precise. It can mean the forms of traditional behavior which are characteristic of a given society, or of a group of societies, or of a certain race, or of a certain area, or of a certain period of time" (Mead, 1937, p. 17); "[c]ulture…consists of regular occurrences in the humanly created world, in the schemas people share as a result of these, and in the interactions between these schemas and this world" (Strauss & Quinn, 1997, p. 7). In an attempt to summarize the different positions Berry et al. (2011) characterized culture as "the shared way of life of a group of people" (p. 4).

However, some scholars believe that culture does not constitute a discrete entity (Gasquoine, 1999; Hermans & Kempen, 1998). Due to the increasing interconnection in a globalized world, Hermans & Kempen (1998) argue that "the concept of independent, coherent, and stable cultures becomes increasingly irrelevant" (p. 1111). The globalization phenomenon is responsible for a large interaction between different cultures leading to the exchange of different economic, political, social, technological, cultural, and ecological elements. This phenomenon has been named cultural convergence or cultural hybridization (Pieterse, 2020). Researchers have developed indexes to measure the degree of globalization of different countries based on these elements (Caselli, 2013), and

DOI: 10.4324/9781003051497-3

even indexes to measure the degree of cultural globalization (Kluver & Fu, 2008). One of the most widely accepted globalization indexes is the KOF Globalization Index (Gygli et al., 2019). This index has a range that runs from 1 to 100. The globalization ranking published in 2019 incorporates cultural globalization elements such as trade in cultural goods, gender parity, trade in personal services, human capital, international trademarks, civil liberties, McDonald's restaurants, and IKEA stores. According to this ranking the first 100 countries (out of 203) have an index above 60 points, i.e., higher than the mid-point of the scale. This list includes "Western" (Switzerland, 91.19 points; Netherlands, 90.71, Belgium, 90,59) as well as "Eastern" (Singapore, 83.62, Malaysia, 81.49, South Korea, 79.29) countries. Therefore, it might be concluded that half the countries are significantly globalized. Moreover, across all countries data indicate a substantial and continuous increase in this index since 1970, especially after 1990 (Gygli et al., 2019).

The idea that cultures are not stable but in constant evolution is also supported by data from anthropological studies. For example, in 1984, Herdt published a book entitled *Ritualized Homosexuality in Melanesia* where he described homosexual behaviors in the rites of sexual initiation in some indigenous tribes (Herdt, 1984). However, by the late 1990s, these sexual initiation behaviors became "vestigial or moribund" (Knauft, 2003, p. 137) as a consequence of the integration of people from these tribes into Western society. It has been estimated that only around 100 tribes still live in isolation nowadays, most of them comprised of a few hundred members (Survival International, 2020). Thus, the idea of stable and completely independent cultures is reduced to this uncertain small number of tribes.

All things considered, it is clear that culture is a broad and complex concept. Because of its breadth and complexity, even its legitimacy has been questioned, leading some anthropologists to suggest that the concept is no longer useful (Abu-Lughod, 1991).

Culture and cross-cultural neuropsychology

Cross-cultural neuropsychology is concerned with the influence of culture on the processes of neuropsychological assessment (including testing of cognitive functions) and rehabilitation. Therefore, it is necessary to determine what specific cultural variables have a significant influence on them. Beyond the considerations of anthropologists about the utility of the concept of culture, it is clear that all previous definitions are of little use for studies in cross-cultural neuropsychology. For example, it is not possible to determine the influence of a "shared way of life" or the "whole complex of traditional behavior" on memory test performance. Contemporary cross-cultural neuropsychology has resorted to concepts such as culture, race, or ethnicity to explain these influences. However, as already noted, these concepts are difficult to operationalize. Gasquoine (1999) suggests that, instead of framing influences on neuropsychological performance in terms of culture, ethnicity or race, given how difficult it is to operationalize these constructs, research should focus on more specific constructs that can be demonstrated to strongly influence neuropsychological performance and are more easily operationalized such as education, language, socio-economic status, and acculturation. Several other authors concur with this view (Byrd et al., 2005; Gasquoine, 1999; Suzuki & Valencia, 1997).

A closer analysis of the concept of race highlights the problems encountered when it has to be operationalized. There is no clear consensus about what race means. As Betancourt & López (1993) state "race is generally defined in terms of physical characteristics, such as skin color, facial features, and hair type, which are common to an inbred, geographically isolated population" (p. 631). Nonetheless, the variability of these characteristics within the same racial group is huge, and even genetic studies cannot define clear boundaries between races (Jorde & Wooding, 2004). Therefore, this concept is highly questioned and seems to be of little use for cross-cultural studies. Similar difficulties arise with the concept of ethnicity which is frequently used to group individuals with a common nationality, culture, or language (Betancourt & López, 1993). But consideration of the wide variety of languages within some countries shows the difficulties that the use of this term poses. For example, there are 11 official languages in South Africa, many of which are spoken by different groups within the black South African population. Thus, even when people may share the same nationality (they are all South African), they may belong to different linguistic and cultural groups. Thus, it is difficult to differentiate ethnicities in this case (as well as in other cases). Moreover, many of these groups share the same skin color (so might be considered to be the same "race") but belong to different linguistic or cultural groups. So, should these people be grouped by race, ethnicity or nationality? In fact, none of these "groupings" are specific enough to usefully define a construct likely to have an impact on neuropsychological assessment and rehabilitation process.

Cultural variables with a significant impact on neuropsychological performance

An alternative approach to the operationalization of the cultural variables that may influence neuropsychological performance is to refer to Murdock's Outline of Cultural Materials, which categorizes variations in cultural practices around the world. He placed these cultural practices in 79 categories that Barry (1980) arranged in eight major sections. These categories are shown in Table 2.1.

The categories described in Table 2.1 are broad. Each one is divided into multiple subcategories. For example, the Behavior Processes & Personality category is divided into the following subcategories: Sensation and Perception, Drivers and Emotions, Modification of Behavior, Adjustment Processes, Personality Development, Social Personality, Personality Traits, Personality Disorders, Life History Materials.

TABLE 2.1 Shared cultural categories of Murdock, arranged by Barry

1. **General Characteristics**	5. **Individual and Family Activities**
Methodology	Living Standards and Routines
Geography	Recreation
Human Biology	Fine Arts
Behavior Processes and Personality	Entertainment
Demography	Social Stratification
History and Culture	Interpersonal Relations
Change	Marriage
Language	Family
Communication	Kinship
2. **Food and Clothing**	6. **Community and Government**
Food Quest	Community
Food Processing	Territorial Organization
Food Consumption	State
Drink, Drugs and Indulgence	Government Activities
Clothing	Political and Sanctions
Adornment	Law
3. **Housing and Technology**	Offenses and Sanctions
Exploitative Activities	Justice
Processing of Basic Materials	War
Building and Construction	7. **Welfare, Religion and Science**
Structures	Social Problems
Settlements	Health and Welfare
Energy and Power	Sickness
Machines	Death
4. **Economy and Transport**	Religious Beliefs
Property	Ecclesiastical Organization
Exchange	Numbers and Measures
Marketing	Ideas About Nature and Man
Finance	8. **Sex and the Life Cycle**
Labor	Sex
Business and Industrial	Reproduction
Organization	Infancy and Childhood
Travel and Transportation	Socialization
	Education
	Adolescence, Adulthood, Old Age

A vast amount of information using this categorization has been compiled in an ethnographic archive known as the Human Relations Area Files (HRAF). This archive provides a description of each category and the possibility of searching information on these categories across different cultures.

These cultural practices might be more easily operationalized than the broader concepts of culture reviewed above. In this way, the influence of each one of the cultural practices on neuropsychological practice (assessment and rehabilitation) might be studied. There is probably little or no information about the relationship between most of these aspects of culture and brain/cognitive functions and their impact on neuropsychological practice. There are reasons to think that some of these cultural factors will not have a significant influence on neuropsychological practice. That is the case for the "objective" or "culture 1" aspects (for example, clothing, adornment or energy, and power) (Hunt, 2007). However, there are sound arguments to study the relationship between some of these cultural practices and neuropsychological assessment and rehabilitation. They are related to the "subjective" or "culture 2" aspects (for example education, language or numbers and measures).

Using this list of cultural practices to guide our future research on the influence of culture on neuropsychological practice may have several advantages: (1) the broad concept, "culture", can be partitioned into specific cultural variables that can be operationalized; (2) the list is comprehensive and considers a large number of variables; (3) the descriptions of each category might be adopted in order to provide a common terminology across the different studies.

In an attempt to determine the possible influence of these aspects of culture on neuropsychological practice this author has selected some of them and organized them into two groups: those which might affect testing and those that probably have little effect on cognitive testing but can definitely affect the wider assessment process, considering that assessment is a broader concept that encompasses testing (see Table 2.2).

Describing the current data available on the relationship between each one of these cultural practices and neuropsychological practice would take a series of

TABLE 2.2 Cultural practices potentially affecting neuropsychological testing and assessment

Impact on Test Performance	*Impact on Wider Aspects of Assessment*
1. Education	1. Interpersonal Relations
2. Language	2. Behavior Processes and Personality
3. Numbers and Measures	3. Communication
4. Human Biology	4. Marriage
5. Communication	5. Family
6. Behavior Processes and Personality	6. Kinship
7. Socialization	7. History
	8. Infancy and Childhood
	9. Socialization
	10. Adolescence; Adulthood, Old Age
	11. Social Problems
	12. Health and Welfare
	13. Religious Beliefs
	14. Ideas About Nature and Man
	15. Law
	16. Justice
	17. War

volumes and is far beyond the scope of this chapter. The reader familiar with the cross-cultural neuropsychological literature may recognize some of these categories on which abundant information has been published. Such is the case of education, which is also discussed in Chapter 4. However, there are other fields in which there is probably little, if any, research (for example, the influence of Religious Beliefs).

A more detailed inspection of the list demonstrates how little we know about the influence of many cultural practices on neuropsychology, but also shows the myriad of possibilities to develop this field. For example, one of the subcategories of "Socialization" is "Transmission of cultural norms" which is defined as instilling of norms concerning appropriate behavior for children (e.g., rest and sleep routines); initiation of children into social relationships (e.g., teaching to share and take turns, instruction in etiquette and kinship behavior); training in gender-typed attitudes and values (e.g., ladylikeness, manliness); instilling of respect for property; instruction in social, ethical, legal and political norms; techniques for education in morals and character; etcetera.[1] How much do we know about the influence of this variable on the neuropsychological practice? Some texts warn about, for example, the different role of women in some Middle-Eastern cultures as compared to Western cultures and the possibility that these attitudinal and practical cultural differences might affect the neuropsychological assessment. Nevertheless, we lack detailed knowledge of the degree this might affect test performance. Do we know precisely what information will be biased, neglected or hidden if a woman is the patient? Do we know if the testing performance of these women will be affected? Which cognitive tests might be affected by this issue? What specific social attitudes will affect performance? Most reports linking these cultural practices to cognitive performance are anecdotal but little empirical research has been conducted on these topics. Without more specific information the practitioner might over- or under-estimate the importance of these variables leading to incorrect clinical decisions. Rather than assuming the influence of these cultural values on the performance of women, it is important to establish whether/how much specific gender-related attitudes and cultural practices impact performance; if a causal association is established, then in the assessment of an individual it would be necessary to establish the extent to which the cultural practice (however measured) was followed by the person being assessed. These questions are completely separate from any judgment on the moral appropriateness of the cultural practices in question.

This analysis is an example of the limited information that is currently available about this particular variable but also the vast number of possibilities to develop the field.

A new model of research in cross-cultural neuropsychology

Considering the former arguments, a new model of research in cross-cultural neuropsychology might be adopted. Rather than comparing the performance of a sample of "Westerners" (for example, Americans) versus "Easterners" (for example

Japanese) on a specific task, the new model should approach research in a different way. For example, if the dependent variable was the performance on a memory task or, preferably, several memory tasks, and the independent variable was "transmission of cultural norms[2]" the study should include at least two groups in which the existence of large differences in the variable under study is well established. Then, using tests to measure each one of the subcomponents of the variable (appropriate behavior for children, initiation of children into social relationships, training in gender-typed attitudes and values, instilling of respect for property, instruction in social, ethical, legal and political norms, techniques for education in morals and character) a multivariate approach should be used (see Figure 2.1). Thus, the influence of these specific variables on the performance on memory tests might be detailed. This approach is not only useful in studying the influence of cultural practices on cognitive testing but also on other components of neuropsychological assessment or rehabilitation. Instead of including performance on a memory test as

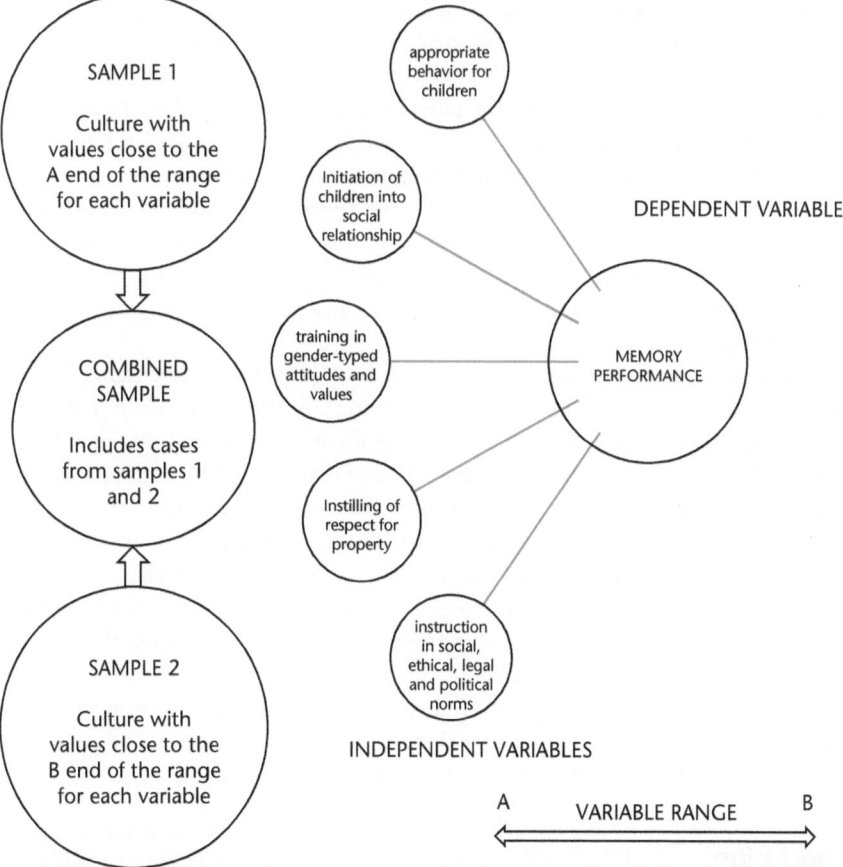

FIGURE 2.1   Example of the study design to investigate the relationship between cultural practices and neuropsychological practices.

the dependent variable other elements that are part of the assessment process might be considered as dependent variables, for example, attitude toward the examination, communication of information or communication difficulties between examiner-examinee. If this approach was replicated with different cultural groups, then more specific information on the influence of these variables on assessment processes might be obtained, which could help discriminate variables that have a significant influence from those that do not.

This approach would also involve changes in the cross-cultural neuropsychological practices themselves. When faced with a culturally different client a clinician should test variables known to have a significant impact in order to understand the potential influence of the factor in this particular situation. The clinician should not assume that someone who is an "Easterner" endorses the cultural values traditionally attributed to Eastern cultures. For example, an American neuropsychologist assessing a Japanese client should not assume a "collectivistic" set of values in this client as the client might be acculturated or simply not identify with those values (this might be particularly true for someone who is an immigrant). The same rationale might be applied in the case of races: a black client should not be assumed to adhere to some cultural values only because of the color of his/her skin. Actually, race is a concept that should be discarded for these purposes because of its lack of utility and high negative social value.

This approach is similar to Ardila's proposal (2007). When discussing the influence of culture on neuropsychological testing he affirmed: "… instead of looking for norms in every existing human group we should try to understand why and how culture impacts cognitive testing, that is, what are the specific cultural variables that may affect the performance in a psychological or neuropsychological test" (p. 39).

Although beyond the scope of this chapter it is worth briefly discussing the implications of this proposal for research on cultural practices and brain anatomy/functioning. Much of the research in this field involves comparing the brain functioning of a sample of "Westerners" versus "Easterners" during the performance of a cognitive task. Although these studies have revealed useful information, this approach is a long way from explaining the complexities of the influence of culture on brain functioning as several intervening variables might be playing an important role in the differences found in these kinds of studies. For example, differences in the activation of some brain regions between "Westerners" and "Easterners" during the performance of a perception task might be due to any of the variables listed on Table 2.2 (for example, language, socialization, human biology, etcetera). Consequently, this approach might help disentangle the influence of these intervening variables in the differences found in brain functioning between different cultural groups.

Conclusion

In summary, the field of cross-cultural neuropsychology has made huge progress in recent decades. This progress has mainly involved the identification and acknowledgment of the influence of cultural variables on neuropsychological assessment and rehabilitation processes. However, a more sophisticated discrimination of the impact of cultural practices is necessary. To that end, the concept of culture needs to be operationalized by defining the specific variables that contribute to the construct of culture. The influence of these variables on neuropsychological assessment rehabilitation should be examined in future studies. Aggregated information from these studies should change the way in which neuropsychological practices with culturally different clients are performed in the future.

Notes

1 This is indeed a broad variable itself that might be divided into several variables.
2 See the HRAF for a detailed description of this construct.

References

Abu-Lughod, L. (1991). Writing against culture. In R. Fox (Ed.), *Recapturing anthropology* (pp. 137–162). Santa Fe, N. Mex.: School of American Research.

Ardila, A. (2007). The impact of culture on neuropsychological test performance. In B.P. Uzzell, M. Pontón, & A. Ardila (Eds.), *International handbook of cross-cultural neuropsychology* (pp. 23–44.). Mahwah, New Jersey: Lawrence Erlbaum Associates.

Barry, H. (1980). Description and uses of the Human Relations Area Files. In H. C., Triandis & J. W., Berry (Eds.), *Handbook of cross-cultural psychology, Vol. II, Methodology,* (pp. 445–478). Boston: Allyn & Bacon.

Berry, J. W., Poortinga, Y. H., Breugelmans, S. M., Chasiotis, A., & Sam, D. L., (2011). *Cross-cultural psychology: Research and applications.* (3 ed.). Cambridge University Press. doi: 10.1017/CBO9780511974274.

Betancourt, H., & López, S.R. (1993). The study of culture, ethnicity, and race in American psychology. *American Psychologist, 48*(6), 629–637. doi: 10.1037/0003-066x.48.6.

Byrd, D.A., Sanchez, D., & Manly, J.J. (2005) Neuropsychological test performance among Caribbean-born and U.S.-born African American elderly: The role of age, education and reading level. *Journal of Clinical and Experimental Neuropsychology, 27*(8), 1056–1069.

Caselli, M. (2013). Nation states, cities, and people: Alternatives methods to measure globalization. *SAGE Open, 3*(4), 1–8. doi: 10.1177/2158244013508417

Gasquoine, P.G. (1999) Variables moderating cultural and ethnic differences in neuropsychological assessment: The case of Hispanic Americans. *The Clinical Neuropsychologist, 13*(3), 376–383.

Gygli, S., Haelg, F., Potrafke, N., & Sturm, J.E. (2019). The KOF globalisation index – revisited. *Review of International Organizations, 14*(3), 543–574, 10.1007/s11558-019-09344-2.

Herdt, G.H. (1984). Ritualized homosexual behavior: An introduction. In G.H. Herdt (Ed.), *Ritualized homosexuality in Melanesia* (pp. 1–81). Berkeley: University of California Press.

Hermans, H.J.M., & Kempen, H.J.G. (1998). Moving cultures: The perilous problems of cultural dichotomies in a globalizing society. *American Psychologist, 53*(10), 1111–1120.

Hunt, R. (2007). *Beyond relativism: Rethinking comparability in cultural anthropology.* Walnut Creek, Calif.: AltaMira Press.

Jorde, L.B., & Wooding, S.P. (2004). Genetic variation, classification and race. *Nature Genetics, 36*(S11). doi: 10.1038/ng1435

Kluver, R., & Fu, W. (2008) Measuring cultural globalization in Southeast Asia. In T. Chong (Ed.), *Globalization and its counter-forces in Southeast Asia* (pp. 335–358). Singapore: Institute of Southeast Asian Studies.

Knauft, B.M. (2003) What ever happened to ritualized homosexuality? Modern sexual subjects in Melanesia and elsewhere. *Annual Review of Sex Research, 14,*137–159.

Kroeber, A.L., & Kluckhohn, C. (1967). *Culture: A critical review of concepts and definitions.* New York: Vintage Books.

Mead, M. (1937). *Cooperation and competition among primitive peoples.* The McGraw-Hill Companies, Inc.

Pieterse, J.N. (2020). *Globalization and culture: global mélange.* Lanham: Rowman & Littlefield.

Strauss, C., & Quinn, N. (1997). *A cognitive theory of cultural meaning.* Cambridge: Cambridge University Press.

Suzuki, L.A., & Valencia, R.R. (1997). Race-ethnicity and measured intelligence: Educational implications. *American Psychologist, 52,* 1103–1114.

Survival International. (n.d.). *Uncontacted tribes.* Retrieved April 6, 2020, from https://www.survivalinternational.org/

3

DEVELOPING CROSS-CULTURAL NEUROPSYCHOLOGY THROUGH THE LENS OF CROSS-CULTURAL COGNITIVE NEUROSCIENCE

Hsu-Wen Huang and Chih-Mao Huang

Psychologists have been investigating individual variations in neurocognitive function since the late 19th century (James, 1890). Neuropsychology, which is a branch of psychology, relies heavily on standardized and validated tests, assessments, and procedures to reliably examine how cognition and behavior are associated with underlying brain function and dysfunction, and scaffolded neural structures. Given the multicultural and multilingual nature of our societies, it has become increasingly important to develop neuropsychological assessments that are valid for culturally heterogeneous groups and that research and clinical neuropsychologists can use. Most studies to date have been conducted by recruiting Western, Educated, Industrialized, Rich, and Democratic (WEIRD; Henrich et al., 2010) participants, with little consideration of neuropsychologically relevant cross-cultural variables. The lack of research focusing on cross-cultural variables in neuropsychology could deprive the field of the knowledge and skills necessary to elucidate the validity and reliability of individual variations in brain-behavior relationships across cultural contexts (Ardila, 1995). Therefore, the influence of cross-cultural variables on cognition, behavior, and brain and related performances is especially vital to consider when using neuropsychological measures and/or norms to assess culturally diverse individuals. However, few attempts have been made to develop viable cross-cultural neuropsychology paradigms by examining culture-related variations in cognition, brain function, and neuroanatomy. Therefore, current studies in the field vastly underestimate the effects of cultural experiences on the assessment and treatment of neurocognitive function.

Although there are a variety of culture-associated dimensions along which cultural groups differ (e.g., power distance), in this chapter, we focus primarily on one dimension: interdependent/collectivistic culture (hypothesized to be

DOI: 10.4324/9781003051497-4

predominant in East Asian cultures such as those in Taiwan, Japan, Korea, Hong Kong, and China), and independent/individualistic culture (hypothesized to be predominant in Western cultures such as those in North America and Western Europe). A large body of psychology and linguistics literature has provided demonstrations that individual differences in information processing result from differences along this cultural dimension. We provide a theoretical framework for understanding and developing cross-cultural neuropsychology. We briefly review empirical evidence from the burgeoning field of cross-cultural cognitive neuroscience, which investigates whether and how sustained exposure to cultural experiences influences the neurobiological basis of human cognition and behavior. We will discuss some cross-cultural neuroimaging studies that address whether neural structures differ across cultural dimensions. We next focus on the consequences of culture-related variations in information processing on neural function, as evidenced by neuroimaging techniques such as functional magnetic resonance imaging (fMRI). Finally, we close with a discussion on the methodological considerations and potential challenges of performing cross-cultural cognitive neuroscience that need to be addressed when developing cross-cultural neuropsychological assessments to evaluate individual behavior-brain associations across cultures.

Theoretical framework of cross-cultural differences

Previous studies in cultural psychology have found that individuals are oriented either toward an interdependent or independent self-construal orientation based on either collectivistic (predominant in East Asian culture) or individualistic (predominant in Western culture) cultural experiences, respectively (Markus & Kitayama, 1991; Triandis, 1995; Triandis et al., 1988); such orientations arise as a result of contrasting cultural values and social beliefs. Following these studies, Nisbett and colleagues proposed a theoretical framework from the cognitive perspective, which has received significant attention for its understanding of the impact of culture on neurocognitive function (Nisbett & Masuda, 2003; Nisbett et al., 2001; Varnum et al., 2010). They proposed that interdependent/collectivistic and independent/individualistic cultures foster dissociable styles of processing information that stem from sustained exposure to specific cultural experiences at the behavioral level. Specifically, East Asian culture has been associated with an emphasis on societal interdependence, collectivistic representation, and a group-based focus on cultural values. East Asians, therefore, tend to view themselves as part of a larger whole, resulting in a bias toward a more holistic style of cognitive processing, in which object and contextual information are jointly processed alongside relations with people and social harmony (i.e., context-inclusive style). In contrast, Western culture tends to favor an individualistic and independent view of the self. This may bias Westerners toward a more analytical style of cognitive processing, reflected in the form of paying more attention to focal objects and organizing information via rules and categories (i.e., object-focused style); it also encourages Westerners to emphasize personal success, independence, and uniqueness (Hong et al., 2001; Nisbett & Masuda, 2003; Nisbett et al., 2001; Oyserman et al., 2002; Varnum et al., 2010).

We are aware of the debate on whether the differences in cognition and social behavior between interdependent/collectivistic and independent/individualistic cultures should be treated as two separate dimensions or a single bipolar dimension (Brewer & Chen, 2007; Oyserman et al., 2002), given the adaptive and dynamic natures of individuals across countries/nations that enable them to endorse different cultural values and social beliefs. Therefore, we note that nationality (e.g., United States [U.S.] or China) and/or cultural affiliation (e.g., American or Chinese) may not necessarily be reliable predictors of cultural values and social beliefs (Chiao et al., 2009; Huang et al., 2019; Oyserman et al., 2002). Despite the debate, systematic culture-related variations have been observed between interdependent/collectivistic and independent-individualistic cultures, as proposed by Nisbett and colleagues, with respect to psychosocial processes such as relationality, social judgment, and self-concept (Han & Northoff, 2008; Huang & Park, 2013; Kitayama & Uskul, 2011; Markus & Kitayama, 1991), and neurocognitive functions such as visual perception, memory, attention, reasoning, and decision making (Doole et al., 2015; Goh & Park, 2009; Nisbett & Masuda, 2003; Nisbett et al., 2001; Park & Huang, 2010; Schwartz et al., 2014).

Culture-related and individual differences in brain structure

Advances in non-invasive structural and functional human neuroimaging techniques such as magnetic resonance imaging (MRI) have provided neuropsychologists with new methods by which to examine the influence of the brain and nervous system on human cognition and behaviors. These techniques can complement the conventional methods (e.g., observation, questionnaires, and/or computerized tasks) used to compare healthy and pathological individuals (Bigler, 2015; Bullmore & Bassett, 2011; Lin et al., 2020; Lungu & Bares, 2020; Roalf & Gur, 2017). Research and clinical neuropsychologists have readily used MRI as an *in vivo* measurement tool to investigate both qualitative and quantitative differences in the human nervous system, and to relate brain structure and connectivity to neuropsychological outcomes (Fan et al., 2019; Kennedy & Raz, 2009; Lin et al., 2019, 2021; Park & Reuter-Lorenz, 2009; Raz et al., 2005; Rodrigue & Raz, 2004). In this section, we review a group of articles that examined the culture-related and individual differences in brain structure, as measured using MRI.

There is a wealth of behavioral and neuroimaging evidence to show that behavioral practices, physical/cognitive training, and prolonged exposure to external experiences shape human brain structure and function (e.g., Chen et al., 2019; Erickson et al., 2011). This suggests that the human brain is adaptive and dynamic. However, we do not know whether the capacity for plasticity of the regional brain volume and cortical thickness can be modulated by culture-related experiences.

To date, only a few cross-cultural and/or cross-national human neuroimaging studies have focused on structural differences in the brains of individuals belonging to East Asian and Western cultures. By harnessing the technique of structural MRI, Zilles et al. (2001) first compared the gross size and shape of the brains of Japanese and European individuals. They demonstrated that the Japanese participants had relatively shorter and wider brains than their European counterparts, with an inter-subject variability distribution seen both across and within cultural samples (Zilles, et al., 2001). Chee et al. (2010) examined a large sample of structural MRI brain images obtained from non-Asian Americans and Singaporean Chinese individuals, collected across two study sites (Singapore and Illinois, U.S.) using identical MR scanner models and the same type of imaging coil. The cortical thickness measurements demonstrated that the non-Asian American participants had thicker cortices than the Singaporean Chinese participants in the frontal, parietal, and temporal polymodal association regions, whereas the Singaporean Chinese participants had higher thicknesses in the left inferior temporal region (Chee et al., 2010). Notably, in Chee et al. (2010), all of the participants underwent a battery of neurocognitive tests (e.g., processing speed, working memory) and scored a minimum of 26 on the Mini-Mental State Examination (MMSE; Folstein et al., 1975), allowing the researchers to ensure well-matched neurocognitive functioning between the two cultural groups. Such MRI findings of cultural differences in brain structure have been reproduced by more recent studies. Tang et al. (2018) compared brain MRI images obtained from male Caucasian individuals, selected from the Human Connectome Project database, with those from male Chinese individuals recruited from the local community in China to explore cultural differences in brain volume and cortical thickness. The results clearly showed that the male Chinese participants had larger brain volumes and cortical thickness in the temporal cortex and cingulate regions, whereas the male Caucasian participants had larger brain volumes and thicker cortical gray matter in the frontal and parietal regions (Tang et al., 2018). Huang et al., (2019) found similar results between East Asians (i.e., Taiwanese) and Westerners (i.e., participants from Europe, U.S., and Australia), in that the Western participants had larger cortical volumes in the fronto-parietal networks, whereas the East Asian participants had larger regional volumes in the temporal and occipital regions (Huang et al., 2019).

In light of the knowledge that nationality (e.g., U.S. or China) and/or cultural affiliation (e.g., American or Chinese) may not reliably reflect cultural groups (Chiao et al., 2009; Oyserman et al., 2002), some structural MRI studies have used self-reported measures of cultural values to examine cultural influences on brain structure, instead of directly comparing individuals belonging to East Asian and Western cultures (Huang et al., 2019; Kitayama et al., 2017; Wang et al., 2017). For example, Wang et al. (2017) collected structural MRI data from a large sample of young Chinese adults to assess individual differences in the orientations of independence-interdependence cultural values. These differences were assessed by administering self-report questionnaires such as the Self-Construal Scale (SCS,

Singelis, 1994). The MRI results of the whole-brain analysis demonstrated that the independence-interdependence score was associated with a larger brain volume, particularly in the right dorsolateral prefrontal and right rostrolateral prefrontal regions, which are thought to be involved in processing self-related information (Wang et al., 2017). Similar results have been reported by Kitayama et al. (2017), who performed structural MRI scans only on young Japanese adults with diverse orientations of independence-interdependence cultural values. They found a significant association between independent-interdependent self-construals and the volume of the bilateral orbitofrontal cortex, with higher interdependent self-construal scores related to larger brain volumes of the orbitofrontal cortex (Kitayama et al., 2017). However, a recent study conducted by Huang et al. (2019), in which East Asian and Western participants were pooled, did not find any significant associations between the orientation of independence-interdependence cultural values and regional brain volumes (Huang et al., 2019).

With respect to the interactions between cultural groups and language systems, an early MRI study examined whether different cultural groups and language systems could interactively mold the structure of the human brain. Morphological brain analyses showed that, when comparing the brains of Chinese-speaking East Asian and English-speaking Caucasian American participants, the regional volumes of the frontal, temporal, and parietal areas were larger in the East Asian participants. These findings suggest that neuroanatomical variations between the two cultural groups are attributable to the linguistic characteristics of Chinese (Kochunov et al., 2003). Similar conclusions were made by Green et al. (2007), who compared the MRI brain images of monolingual (i.e., English) and multilingual (i.e., Chinese and English) individuals. They reported higher brain densities in the frontal and temporal regions of Chinese multilingual speakers (Green et al., 2007), suggesting that the language system may modulate and/or interact with brain structure in the context of culture-specific experiences.

In sum, several cross-cultural neuroimaging studies that compared the regional brain structures between East Asians and Westerners have shown significant culture-associated variations in brain volume and cortical thickness. However, the association between individual differences in independent-interdependent orientation and brain structure was inconclusive, implying that more research is needed to resolve this aspect. Finally, we note several discrepancies between the different MRI readouts and cultural samples, which may be driven by genetic diversity, environmental biases, and/or gene-environment interactions across multiple time scales (Chee et al., 2010; Chiao, et al, 2013; Goh & Huang, 2012; Huang et al., 2019; Yu et al., 2018).

FMRI data demonstrating cultural influences on brain function

As an applied psychological science, neuropsychology is concerned with the behavioral expression of brain function and dysfunction both in healthy people and in heterogeneous patient groups. The development of fMRI techniques has cemented the possibility of studying behavior-brain associations through *in vivo* measurements of the neurophysiological substrates of brain function/dysfunction and the neurocognitive mechanisms of brain activities (Bennett et al., 2016; Huang et al., 2012, 2019; Roalf & Gur, 2017; Sutterer & Tranel, 2017). fMRI has several variants (e.g., resting-state and task-evoked fMRI), but most are sensitive to changes in blood oxygenation level-dependent (BOLD) signals, which reflect the magnetic properties of blood that is oxygenated to different degrees, induced by the activation of neurons thought to be associated with specific cognitive functions, behavioral performances, or affective processes (Huettel et al., 2014; Kwong et al., 1992; Ogawa et al., 1990). Below, we briefly introduce some fMRI findings on differences in cultural influences on brain activation between East Asian and Western individuals.

The cultural dichotomy in information processing (i.e., context-inclusive styles for East Asian and object-focused styles for Western cultures) presents a convenient path along which to explore the manifestations of such culture-related variations in the brain (for a review, see Goh & Park, 2009; Han & Northoff, 2008; Park & Huang, 2010). Although some fMRI studies have examined the neural signatures of culture-related differences in cognition, their findings suggest that the brains of individuals from different cultures employ different cognitive strategies to process information, particularly in the ventral-visual (Goh et al., 2007, 2011; Gutchess et al., 2006; Jenkins et al., 2010) and fronto-parietal (Chiao et al, 2009; Goh et al., 2013; Gutchess et al., 2010; Hedden et al., 2008; Pornpattananangkul et al., 2016) regions. For example, in a cross-cultural fMRI study, Gutchess et al., (2006) instructed East Asian and Western participants to passively view pictures of salient objects (e.g., elephant), of scenic backgrounds with no salient objects embedded (e.g., jungle), and of salient objects against scenic backgrounds (e.g., an elephant in a jungle). In keeping with the behavioral findings, the fMRI results showed that the Western participants had higher levels of activation in their object-processing regions, including the bilateral middle temporal gyrus, left superior parietal gyrus, and right superior temporal gyrus, although no significant activation differences were observed in the context-processing regions of the East Asian participants (Gutchess et al., 2006). Similar cross-cultural fMRI results have been reported by Goh et al. (2010) in the context of viewing face and house stimuli, and by Jenkins et al. (2010) in the context of viewing congruent (e.g., a cow in the farm) and incongruent (e.g., a cow in a kitchen) scenes. These findings suggest that the cultural group bias that operates on the ventral-visual cortex is associated with visual perception and attention processing, as indicated by previous eye-movement and behavioral data.

A few studies have examined whether and how cultural experiences influence the neural functioning of executive control. For example, Pornpattananangkul et al. (2016) directed Caucasian American, Japanese American, and native Japanese participants to perform English and Japanese versions of the go/no-go task. Although there were no cultural differences in the behavioral responses, the researchers found that the levels of brain activation in the fronto-parietal regions were modulated by cultural affiliations, with the greater inferior frontal gyrus (IFG) being more active in the native Japanese participants than in the Caucasian American and Japanese American participants; in contrast, no systematic differences in IFG activation were observed between the Japanese American and Caucasian American participants (Pornpattananangkul et al., 2016). Another study, conducted by Hedden et al. (2008), that used a modified version of the frame-line test adapted from Kitayama et al. (2003) found that Western participants showed higher levels of fronto-parietal activation during relative judgments than during absolute line judgments. In contrast, East Asian participants showed significantly larger fronto-parietal responses during absolute judgments than during relative judgments (Hedden et al., 2008). These findings suggest that individuals show greater fronto-parietal activation when processing non-culturally preferred cognitive tasks, which probably require more resources to achieve the respective goals.

With respect to culture-related differences in language processing, a few fMRI studies on reading-impaired Chinese children (i.e., dyslexia) have shown that culture constrains the fundamental pathophysiology of dyslexia (Siok et al., 2004), whereas other cross-language fMRI studies have shown universal neural activation to be associated with proficient language processing across a variety of language systems (Chee et al., 2000; Rueckl et al., 2015). For example, by contrasting four distinct languages – Spanish, English, Hebrew, and Chinese – Rueckl et al. (2015) instructed native speakers of each language from different countries (i.e., Spain, U.S., Israel, and Taiwan) to make semantic judgments (living/non-living) of print or auditory stimuli. They found an extensive convergence of printed and spoken language processing in the fronto-parietal and middle temporal regions, which are known to be associated with semantic processing (Rueckl et al., 2015), suggesting the use of common brain systems for language processing across cultures.

There is currently substantial cross-cultural behavioral evidence to show that Westerners are more likely to organize information via rules and categories, whereas East Asians are more likely to focus on contextual information and prioritize relational information over categorical information. In light of this evidence, Gutchess et al., (2010) examined whether cultural groups show differential brain activation when performing functional-relationship (e.g., cow-grass) and categorical-relationship (e.g., cow-chicken) semantic judgment tasks. They reported that the East Asian participants showed higher levels of activation in the fronto-parietal brain regions, which likely mediate controlled attentional processes, whereas the American participants showed higher levels of activation in the temporal regions, likely in response to conflicting semantic information. These findings indicate the presence of cultural variations in the strategies employed to resolve conflicts between competing semantic judgments (Gutchess et al., 2010). Taken together, these compelling fMRI findings suggest that the brains of individuals from different cultural groups use different cognitive strategies to process culturally preferred information. Importantly, these strategies are associated with a variety of cognitive domains both in the ventral-visual and fronto-parietal brain regions.

Methodological considerations when examining cross-cultural differences in neurocognitive function

The cross-cultural cognitive neuroscience research summarized above provides theoretical, methodological, and empirical foundations for the understanding and evaluation of individual variations in neurocognitive function across cultural contexts. Given that the societies in East Asia (e.g., China) and the West (e.g., U.S.) are multicultural, it is vital to develop neuropsychological assessments that are valid for culturally heterogeneous groups. Therefore, appropriate qualitative modifications to traditional measures and norms have to be made when conventional methods fail to accurately describe individual behaviors in cross-cultural settings.

The focus here is on the methodological considerations salient to the development of culturally relevant approaches to neuropsychological assessment. First, the impacts of subtle cultural factors on objective measures of performance used to elucidate individual variations in brain-behavior relationships should be considered. For example, Hedden et al., (2002) clearly demonstrated that when numerical tests to assess processing speed (measured by digit comparison) and working memory (measured by backward digit span) were performed, young Chinese participants performed better than young American participants. However, such culture-related differences disappeared when spatial tasks were used to measure processing speed (measured by pattern comparison) and working memory (measured by backward Corsi blocks). The misleading conclusion of a greater neurocognitive performance by Chinese participants was, therefore, valid only when numerical tests were used, making these tests inappropriate indicators for neuropsychological assessments across cultural contexts; these tests were rendered unsuitable due to the facilitatory effects of the linguistic properties of the Chinese language on numerical representation and processing, which had been overlooked earlier (Cheung & Kemper, 1993, 1994; Huang et al., 2021; Nuerk et al., 2005). Specifically, the structures of Chinese number words clearly map onto the place-value features of the Arabic numeral system, which is consistent with the conventional base-ten numeration system. For example, the number "11" is lexicalized as "shi2 yi1", which literally means "ten one" in Chinese, whereas it is lexicalized as "eleven" in English, lacking a clear mapping onto "ten" and "one". Furthermore, the phonological characteristic of Chinese number names is much simpler than those in English. In Chinese, all single-digit numbers have single syllables, whereas in English the name for "7" contains two syllables. These subtle cultural factors and linguistic properties, therefore, facilitate Arabic number comparisons, counting, and general numerical performance using Arabic notations in Chinese speakers, biasing the neurocognitive assessments of such individuals (Huang et al., 2021; Nuerk et al., 2005; Tzeng & Wang, 1983). Therefore, we infer that unequivocal conclusions with respect to cross-cultural differences in cognition can be made only when the appropriate experimental stimuli are selected for studies and assessments.

A second issue to keep in mind when conducting neuropsychological assessments and neuroimaging studies across cultural groups is the importance of determining the validity and reliability of norms for a comprehensive set of neuropsychological tests (e.g., language abilities and verbal knowledge) for different cultural groups stratified by age, gender, and educational level (Bates et al., 2003; Li et al., 2021; Yoon et al., 2004a, 2004b). For example, Yoon et al. (2004b) conducted a normative measures test for a pictorial dataset (i.e., 260 line drawings of objects) that was developed and used for neuropsychological investigations in Western societies. Responses for name agreement, concept agreement, and familiarity measures were separately obtained from each cultural group recruited from China and the U.S. The study reported significant culture-related differences in the percentage of name agreement responses and the number of distinct name responses across all culture-by-age groups (Yoon et al., 2004b). Moreover, semantic categorization is sensitive to the identification of the selective impairment of domain-specific semantic knowledge, which is attributable to brain damage that can cause differences in language processing with respect to inanimate versus animate (Caramazza & Shelton, 1998) and concrete versus abstract objects (Catricalà et al., 2014; Huang & Federmeier, 2015). In light of this evidence, Yoon et al. (2004a) performed a semantic category norming study using different language materials to examine whether the language database available at the time was comparable in terms of semantic categorization across cultures (i.e., Chinese and American participants). In this cross-cultural, cross-linguistic measure (i.e., Chinese versus English), the unusually complex processes of coding, categorization, and cross-translation were taken into account to develop a standardized, cross-referenced, and language-based dataset across cultural contexts. The study reported that only six of the 105 categories tested showed high-affinity scores (≥0.90) when the two cultural groups were compared, suggesting a substantial difference in category norms between cultures (Yoon et al., 2004a). Similar work on category norms has examined individual differences in cognition across subgroups of Chinese cultures, with individuals who may be from distinct cultures (e.g., China, Taiwan, Hong Kong) and whose primary language/dialect may be varied (i.e., Mandarin, Cantonese) (Li et al., 2021).

Taken together, the successful outcomes of these norming studies highlight the necessity for and research potential of a database that provides a cross-culturally valid and sensitive approach to the investigation of individual variations in neurocognitive functions. We, therefore, suggest that current neuropsychological stimuli, norms, and tests be systematically reexamined for cross-cultural propriety.

Conclusions and future directions

There is sufficient evidence to show that East Asian and Western cultures present dissociable styles of processing information (i.e., context-inclusive style for East Asians and object-focused style for Westerners) that modulate neurocognitive processes, influence neural functions, and mold brain structures due to sustained exposure to interdependent/collectivistic or independent/individualistic cultural experiences, respectively. The empirical findings of cross-cultural cognitive neuroscience studies suggest that a fusion between the culturally-invariant and culturally-specific characteristics of neurocognitive functions should be adopted as neuropsychology moves forward as a culturally sensitive science. Moreover, the rapid evolution of neuroimaging techniques will enable us to develop neuropsychological methods and assessments that successfully integrate traditional approaches with more ecologically valid, culturally relevant, and functional approaches to documenting and investigating brain-behavior associations across a variety of cultural societies to which North American tests and norms may not apply. Finally, the primary direction for the development of broader frameworks of cross-cultural neuropsychology should be one in which the influence of well-documented variables such as age (Goh & Huang, 2012; Park, 2008; Park & Gutchess, 2002, 2006), gender, educational level (Huang et al., 2019; Huang & Huang, 2019; Lin et al., 2020), bilingualism, and occupational complexity and their interactions with cultural backgrounds are considered, all of which can affect neuropsychological performance and brain-behavior associations in both healthy and pathological populations.

Acknowledgments

This work was supported by a Strategic Research Grant (7200538), by the Hong Kong Institute for Advanced Study (9360157), by the City University of Hong Kong, and by the Ministry of Science and Technology (103-2420-H-009-006-MY2; 105-2420-H-009-001-MY2; 107-2410-H-009-028-MY3) of Taiwan. H. W. Huang and C. M. Huang would like to thank Shih-Ping Huang for his company and indispensable support during the COVID-19 quarantine period.

References

Ardila, A. (1995). Directions of research in cross-cultural neuropsychology. *Journal of Clinical and Experimental Neuropsychology, 17*(1), 143–150.

Bates, E., D'Amico, S., Jacobsen, T., Székely, A., Andonova, E., Devescovi, A., Herron, D., Lu, C.C., Pechmann, T., Pleh, C., Wicha, N., Federmeier, K., Gerdjikova, I., Gutierrez, G., Hung, D., Hsu, J., Iyer, G., Kohnert, K., Mehotcheva, T., Orozco-Figueroa, A., Tzeng, A., & Tzeng, O. (2003). Timed picture naming in seven languages. *Psychonomic Bulletin and Review, 10*(2), 344–380.

Bennett, M.R., Hatton, S., Hermens, D.F., & Lagopoulos, J. (2016). Behavior, neuropsychology and fMRI. *Progress in Neurobiology, 145,* 1–25.

Brewer, M.B., & Chen, Y.-R. (2007). Where (who) are collectives in collectivism? Toward conceptual clarification of individualism and collectivism. *Psychological Review, 114*(1), 133–151.

Bigler, E.D. (2015). Structural image analysis of the brain in neuropsychology using magnetic resonance imaging (MRI) techniques. *Neuropsychology Review, 25*(3), 224–249.

Bullmore, E.T., & Bassett, D.S. (2011). Brain graphs: Graphical models of the human brain connectome. *Annual Review of Clinical Psychology, 7*(1), 113–140.

Caramazza, A., & Shelton, J.R. (1998). Domain-specific knowledge systems in the brain: The animate-inanimate distinction. *Journal of Cognitive Neuroscience, 10*(1), 1–34.

Catricalà, E., Della Rosa, P.A., Plebani, V., Vigliocco, G., & Cappa, S.F. (2014). Abstract and concrete categories? Evidences from neurodegenerative diseases. *Neuropsychologia, 64*, 271–281.

Chee, M.W.L., Weekes, B., Lee, K.M., Soon, C.S., Schreiber, A., Hoon, J.J., & Chee, M. (2000). Overlap and dissociation of semantic processing of Chinese characters, English words, and pictures: Evidence from fMRI. *Neuroimage, 12*(4), 392–403.

Chee, M.W.L., Zheng, H., Goh, J.O.S., & Park, D. (2010). Brain structure in young and old East Asians and Westerners: Comparisons of structural volume and cortical thickness. *Journal of Cognitive Neuroscience, 23*(5), 1065–1079.

Chen, F.T., Chen, Y.P., Schneider, S. Kao, S.C., Huang C.M., & Chang, Y.K. (2019). Effects of exercise modes on neural processing of working memory in late middle-aged adults: An fMRI study. *Frontiers in Aging Neuroscience, 11*, 224.

Cheung, H., & Kemper, S. (1993). Recall and articulation of English and Chinese words by Chinese-English bilinguals. *Memory and Cognition, 21*(5), 666–670.

Cheung, H., & Kemper, S. (1994). Recall and articulation of English and Chinese words under memory preload conditions. *Language and Speech, 37*(2), 147–161.

Chiao, J.Y., Cheon, B.K., Pornpattanangkul, N., Mrazek, A.J., & Blizinsky, K.D. (2013). Cultural neuroscience: Progress and promise. *Psychological Inquiry, 24*(1), 1–19.

Chiao, J.Y., Harada, T., Komeda, H., Li, Z., Mano, Y., Saito, D., Parrish, T.B., Sadato, N., & Iidaka, T. (2009). Neural basis of individualistic and collectivistic views of self. *Human Brain Mapping, 30*(9), 2813–2820.

Doole, R. Chan, M.Y., & Huang, C.M. (2015). Intercultural relations and the perceptual brain: A cognitive neuroscience perspective. In J.E. Warnick & D. Landis (Eds.), *Neuroscience in intercultural contexts* (pp. 203–214). Springer.

Erickson, K.I., Voss, M.W., Prakash, R.S., Basak, C., Szabo, A., Chaddock, L., Kim, J.S., Heo, S., Alves, H., White, S.M., & Wojcicki, T.R. (2011). Exercise training increases size of hippocampus and improves memory. *Proceedings of the National Academy of Sciences of the United States of America, 108*(7), 3017–3022.

Fan, Y.T., Fang, Y.W., Chen, Y.P., Leshikar, E.D., Lin, C.P., Tzeng, O.J.L., Huang, H.W., & Huang, C.M. (2019). Aging, cognition, and the brain: Effects of age-related variation in white matter integrity on neuropsychological function. *Aging and Mental Health, 23*(7), 831–839.

Folstein, M.F., Folstein, S.E., & McHugh, P.R. (1975). "Mini-mental state." A practical method for grading the cognitive state of patients for the clinician. *Journal of Psychiatric Research, 12*(3), 189–198.

Goh, J.O., Chee, M.W., Tan, J.C., Venkatraman, V., Hebrank, A., Leshikar, E.D., Jenkins, L., Sutton, B.P., Gutchess, A.H., & Park, D.C. (2007). Age and culture modulate object processing and object-scene binding in the ventral visual area. *Cognitive, Affective, and Behavioral Neuroscience, 7*(1), 44–52.

Goh, J. O., Leshikar, E. D., Sutton, B. P., Tan, J. C., Sim, S. K., Hebrank, A. C., & Park, D. C. (2010). Culture differences in neural processing of faces and houses in the ventral visual cortex. *Social Cognitive and Affective Neuroscience, 5*(2–3), 227–235. 10.1093/scan/nsq060.

Goh, J.O.S., Hebrank, A.C., Sutton, B.P., Chee, M.W.L., Sim, S.K.Y., & Park, D.C. (2013). Culture-related differences in default network activity during visuo-spatial judgments. *Social Cognitive and Affective Neuroscience, 8*(2), 134–142.

Goh, J.O., & Huang, C.M. (2012). Images of the cognitive brain across age and culture. InP.. Bright (Ed.), *Neuroimaging-cognitive and clinical neuroscience* (pp. 17–46). InTech.

Goh, J.O., & Park, D.C. (2009). Culture sculpts the perceptual brain. *Progress in Brain Research, 178,* 95–111.

Green, D.W., Crinion, J., & Price, C.J. (2007). Exploring cross-linguistic vocabulary effects on brain structures using voxel-based morphometry. *Bilingualism: Language and Cognition, 10*(2), 189–199.

Gutchess, A.H., Hedden, T., Ketay, S., Aron, A., & Gabrieli, J.D.E. (2010). Neural differences in the processing of semantic relationships across cultures. *Social Cognitive and Affective Neuroscience, 5*(2-3), 254–263.

Gutchess, A.H., Welsh, R.C., Boduroğlu, A., et al. Cultural differences in neural function associated with object processing. *Cognitive, Affective, & Behavioral Neuroscience 6,* 102–109 (2006). 10.3758/CABN.6.2.102.

Han, S., & Northoff, G. (2008). Culture-sensitive neural substrates of human cognition: A transcultural neuroimaging approach. *Nature Reviews Neuroscience, 9*(8), 646–654.

Hedden, T., Ketay, S., Aron, A., Markus, H.R., & Gabrieli, J.D. (2008). Cultural influences on neural substrates of attentional control. *Psychological Science, 19*(1), 12–17.

Hedden, T., Park, D.C., Nisbett, R., Ji, L.-J., Jing, Q., & Jiao, S. (2002). Cultural variation in verbal versus spatial neuropsychological function across the life span. Neuropsychology, 16(1), 65–73. 10.1037/0894-4105.16.1.65.

Henrich, J., Heine, S.J., & Norenzayan, A. (2010). The weirdest people in the world? *Behavioral and Brain Science*s, *33,* 61–83.

Hong, Y., Ip, G., Chiu, C., Morris, M.W., & Menon, T. (2001). Cultural identity and dynamic construction of the self: Collective duties and individual rights in Chinese and American cultures. *Social Cognition, 19,* 251–268.

Huang, C.M., Doole, R., Wu, C.W., Huang, H.W., & Chiao, Y.P. (2019). Culture-related and individual differences in regional brain volumes: A cross-cultural voxel-based morphometry study. *Frontiers in Human Neuroscience. 13,* 313.

Huang, C.M., Fan, Y.T., Lee, S.H., Liu, H.L., Lin, C.M., & Lee, T.M.C. (2019). Cognitive reserve-mediated neural modulation of emotional control and regulation in people with late-life depression. *Social Cognitive and Affective Neuroscience. 14*(8), 849–860.

Huang, C.M., & Huang, H.W. (2019). Aging, neurocognitive reserve, and the healthy brain. In *Psychology of learning and motivation* (*Vol 71,* pp. 175–213). Academic Press.

Huang, C.M. & Park, D.C. (2013). Cultural influences on Facebook photographs. *International Journal of Psychology, 48*(3), 334–343.

Huang, C.M., Polk, T.A., Goh, J.O., & Park, D.C. (2012). Both left and right posterior parietal activations contribute to compensatory processes in normal aging. *Neuropsychologia, 50*(1), 55–66.

Huang, F.Y., Hsu, A.L., Hsu, L.M., Tsai, J.S., Huang, C.M., Chao, Y.P., Hwang, T.J., Wu, C.W.W. (2019). Mindfulness improves emotion regulation and executive control on bereaved individuals: An fMRI study. *Frontiers in Human Neuroscience. 12,* 541.

Huang, H.W. (2016). Morphological processing of compounds: Neurolingustic studies. In Sybesma R. (Ed.), *Encyclopedia of Chinese language and linguistics* (*Vol. 3,* pp. 100–104). Brill.

Huang, H.W., Federmeier K.D. (2015). Imaginative language: What event-related potentials have revealed about the nature and source of concreteness effects. *Language and Linguistics, 16*(4), 503–515.

Huang, H.W., Nascimben, M., Wang, Y.Y., Fong, D.Y., Tzeng, O.J.L., & Huang, C.M. (2021). Which digit is larger? Brain responses to number and size interactions in a numerical Stroop task. *Psychophysiology, 58*(3), e13744.

Huettel, S.A., Song, A.W., & McCarthy, G. (2014). *Functional magnetic resonance imaging.* Sinauer Associates.

James, W. (1890). *Principles of psychology.* Holt & Co.

Jenkins, L.J., Yang, Y.-J., Goh, J., Hong, Y.-Y., & Park, D.C. (2010). Cultural differences in the lateral occipital complex while viewing incongruent scenes. *Social Cognitive and Affective Neuroscience, 5*(2-3), 236–241.

Kennedy, K.M., Raz, N. (2009). Aging white matter and cognition: Differential effects of regional variations in diffusion properties on memory, executive functions, and speed. *Neuropsychologia, 47*(3), 916–927.

Kitayama, S., Duffy, S., Kawamura, T., & Larsen, J.T. (2003). Perceiving an object and its context in different cultures: A cultural look at new look. *Psychological Science, 14*(3), 201–206.

Kitayama, S., & Uskul, A.K. (2011). Culture, mind, and the brain: Current evidence and future directions. *Annual Review of Psychology, 62*, 419–449.

Kitayama, S., Yanagisawa, K., Ito, A., Ueda, R., Uchida, Y., & Abe, N. (2017). Reduced orbitofrontal cortical volume is associated with interdependent self-construal. *Proceedings of the National Academy of Sciences of the United States of America, 114*(30), 7969–7974.

Kochunov, P., Fox, P., Lancaster, J., Tan, L.H., Amunts, K., Zilles, K., Mazziotta, J., & Gao, J.H. (2003). Localized morphological brain differences between English-speaking Caucasians and Chinese-speaking Asians: New evidence of anatomical plasticity. *Neuroreport, 14*(7), 961–964.

Kwong, K.K., Belliveau, J.W., Chesler, D.A., Goldberg, I.E., Weisskoff, R.M., Poncelet, B.P., Kennedy D.N., Hoppel, B.E., Cohen, M.S., & Turner, R. (1992). Dynamic magnetic resonance imaging of human brain activity during primary sensory stimulation. *Proceedings of the National Academy of Sciences of the United States of America, 89*(12), 5675–5679.

Lezak, M.D. (1995). *Neuropsychological assessment.* Oxford University Press.

Li, B., Lin, Q., Mak, H., Tzeng, O.J.L., Huang, C.M., Huang, H.W. (2021). Category exemplar production norms for Hong Kong Cantonese: Instance probabilities and word familiarity. *Frontiers in Psychology.* doi: 10.3389/fpsyg.2021.657706.

Lin, C.M., Huang, C.M., Fan, Y.T., Liu, H.L., Cheg, Y.L., Aizenstein, H.J., Lee, T.M.C., & Lee, S.H. (2020). Cognitive reserve moderates effects of white matter hyperintensity on depressive symptoms and cognitive function in late-life depression. *Frontiers in Psychiatry. 11*, 249.

Lin, C.M., Huang, C.M., Karim H.T., Liu, H.L., Lee, T.M.C., Wu, C.W.W., Toh, C.H., Tsai, Y.F., Yen, T.H., & Lee, S.H. (2021). Greater white matter hyperintensities and the association with executive function in suicide attempters with late-life depression. *Neurobiology of Aging. 103*, 60–67.

Lin, C.M., Lee, S.H., Huang, C.M., Chen, G.Y., Ho, P.S., Liu, H.L., Chen, Y.L., Lee, T.M.C., & Wu, S.C. (2019). Increased brain entropy of resting-state fMRI mediates the relationship of depression severity and mental health-related quality of life in late-life depressed elderly. *Journal of Affective Disorders. 250*, 270–277.

Lungu, O., & Bares, M. (2020). Editorial: Neuropsychology through the MRI looking glass. *Frontiers in Neurology, 11*(1365), 609897.

Markus, H., & Kitayama, S. (1991). Culture and the self: Implication for cognition, emotion and motivation. *Psychological Review, 98*(2), 224–253.

Nuerk, H.C., Weger, U., & Willmes, K. (2005). Language effects in magnitude comparison: Small, but not irrelevant. *Brain and Language, 92*(3), 262–277.

Nisbett, R.E., & Masuda, T. (2003). Culture and point of view. *Proceedings of the National Academy of Sciences of the United States of America, 100*(19), 11163–11170.

Nisbett, R.E., Peng, K., Choi, I., & Norenzayan, A. (2001). Culture and systems of thought: Holistic versus analytic cognition. *Psychological Review, 108*(2), 291–310.

Ogawa, S., Lee, T.M., Kay, A.R., & Tank, D.W. (1990). Brain magnetic resonance imaging with contrast dependent on blood oxygenation. *Proceedings of the National Academy of Sciences of the United States of America, 87*(24), 9868–9872.

Oyserman, D., Coon, H.M., & Kemmelmeier, M. (2002). Rethinking individualism and collectivism: Evaluation of theoretical assumptions and meta-analyses. *Psychological Bulletin, 128*(1), 3–72.

Patil, A.U., Madathil, D., & Huang, C.M. (2021). Healthy aging alters the functional connectivity of creative cognition in the default mode network and cerebellar network. *Frontiers in Aging Neuroscience, 13*, 31.

Park, D.C. (2008). Developing a cultural cognitive neuroscience of aging. In Hofer, S.M., & Alwin, D.F. (Eds.), *Handbook of Cognitive Aging: Interdisciplinary Perspectives* (pp. 352–367). Sage.

Park, D.C., & Gutchess, A.H. (2002). Aging, cognition, and culture: A neuroscientific perspective. *Neuroscience and Biobehavioral Reviews, 26*(7), 859–867.

Park, D.C., & Gutchess, A.H. (2006). The cognitive neuroscience of aging and culture. *Current Directions in Psychological Science, 15*(3), 105–108.

Park, D.C. & Huang, C.M. (2010). Culture wires the brain: A cognitive neuroscience perspective. *Perspectives on Psychological Science, 5*(4), 391–400.

Park, D.C., Nisbett, R.E., & Hedden, T. (1999). Aging, culture, and cognition. *Journals of Gerontology: Series B. Psychological Sciences and Social Sciences, 54*(2), P75–P84.

Park, D.C., & Reuter-Lorenz, P. (2009). The adaptive brain: Aging and neurocognitive scaffolding. *Annual Review of Psychology, 60*, 173–196.

Pornpattananangkul, N., Hariri, A.R., Harada, T., Mano, Y., Komeda, H., Parrish, T.B., Sadato, N., Iidaka, T., & Chiao, J.Y. (2016). Cultural influences on neural basis of inhibitory control. *Neuroimage, 139*, 114–126.

Raz, N., Lindenberger, U., Rodrigue, K.M., Kennedy, K.M., Head, D., Williamson, A., Dahle, C., Gerstorf, D., & Acker, J.D. (2005). Regional brain changes in aging healthy adults: General trends, individual differences and modifiers. *Cerebral Cortex, 15*(11), 1676–1689.

Roalf, D.R., & Gur, R.C. (2017). Functional brain imaging in neuropsychology over the past 25 years. *Neuropsychology, 31*(8), 954–971.

Rodrigue, K.M., & Raz, N. (2004). Shrinkage of the entorhinal cortex over five years predicts memory performance in healthy adults. *Journal of Neuroscience, 24*(4), 956–963.

Rueckl, J.G., Paz-Alonso, P.M., Molfese, P.J., Kuo, W.-J., Bick, A., Frost, S.J., Hancock, R., Wu, D.H., Mencl, W.E., Duñabeitia, J.A., & Lee, J.R. (2015). Universal brain signature of proficient reading: Evidence from four contrasting languages. *Proceedings of the National Academy of Sciences of the United States of America, 112*(50), 15510–15515.

Schwartz, A.J., Boduroglu, A., & Gutchess, A.H. (2014). Cross-cultural differences in categorical memory errors. *Cognitive Science, 38*(5), 997–1007.

Siok, W.T., Perfetti, C.A., Jin, Z., & Tan, L.H. (2004). Biological abnormality of impaired reading is constrained by culture. *Nature, 431*(7004), 71–76.

Singelis, T. M. (1994). The measurement of independent and interdependent self-construals. *Personality and Social Psychology Bulletin, 20*(5), 580–591.

Sutterer, M.J., & Tranel, D. (2017). Neuropsychology and cognitive neuroscience in the fMRI era: A recapitulation of localizationist and connectionist views. *Neuropsychology, 31*(8), 972–980.

Tang, Y., Zhao, L., Lou, Y., Shi, Y., Fang, R., Lin, X., Liu, S., & Toga, A. (2018). Brain structure differences between Chinese and Caucasian cohorts: A comprehensive morphometry study. *Human Brain Mapping, 39*(5), 2147–2155.

Triandis, H.C. (1995). *Individualism and collectivism.* Westview Press.

Triandis, H.C., Bontempo, R., Villareal, M.J., Asai, M., & Lucca, N. (1988). Individualism and collectivism: Cross-cultural perspectives on self-ingroup relationships. *Journal of Personality and Social Psychology, 54*(2), 323–338.

Tzeng, O.J., & Wang, W.S. (1983). The first two R's. The way different languages reduce speech to script affects how visual information is processed in the brain. *American Scientist, 71*(3), 238–243.

Varnum, M.E.W., Grossmann, I., Kitayama, S., & Nisbett, R.E. (2010). The origin of cultural differences in cognition: Evidence for the social orientation hypothesis. *Current Directions in Psychological Science, 19*(1), 9–13.

Wang, F., Peng, K., Chechlacz, M., Humphreys, G.W., & Sui, J. (2017). The neural basis of independence versus interdependence orientations: A voxel-based morphometric analysis of brain volume. *Psychological Science, 28*(4), 519–529.

Yoon, C., Feinberg, F., Hu, P., Gutchess, A.H., Hedden, T., Chen, H.-Y.M., Jing, Q., Cui, Y., & Park, D.C. (2004a). Category norms as a function of culture and age: Comparisons of item responses to 105 categories by American and Chinese adults. *Psychology and Aging, 19*(3), 379–393.

Yoon, C., Feinberg, F., Luo, T., Hedden, T., Gutchess, A.H., Chen, H.-Y.M., Mikels, J.A., Jiao, S., & Park, D.C. (2004b). A cross-culturally standardized set of pictures for younger and older adults: American and Chinese norms for name agreement, concept agreement, and familiarity. *Behavior Research Methods, Instruments, and Computers, 36*(4), 639–649.

Yu, Q., Abe, N., King, A., Yoon, C., Liberzon, I., & Kitayama, S. (2018). Cultural variation in the gray matter volume of the prefrontal cortex is moderated by the dopamine D4 receptor gene (DRD4). *Cerebral Cortex, 29*(9), 3922–3931.

Zilles, K., Kawashima, R., Dabringhaus, A., Fukuda, H., & Schormann, T. (2001). Hemispheric shape of European and Japanese brains: 3-D MRI analysis of intersubject variability, ethnical, and gender differences. *Neuroimage, 13*(2), 262–271.

4

EDUCATION, THE MOST POWERFUL CULTURAL VARIABLE?

Alberto Luis Fernández

Influence of education

The powerful influence of formal education on neuropsychological test performance is well recognized (Ardila & Rosselli, 2007; Marcopulos et al., 1997). Education, and literacy in particular, affects how examinees approach situations in which they are being tested (Ardila et al., 2010; Grossi et al., 1993; Nell, 2000) and also changes the way that brains function (Castro-Caldas & Reis, 2000). The approach to the testing situation has been referred to as "test-wiseness" (Ardila et al., 2010; Nell, 2000), meaning that the examinee is familiar with the testing situation: being alone in a room with a test administrator, time limits to complete tasks, the expectation of a good performance, etcetera. But individuals who have been to school are not only test-wise but have typically developed specific skills and strategies relevant to tasks commonly used as part of a neuropsychological assessment. The school experience has demonstrated influences on IQ, mathematics, visual perception, semantic and phonological processing, reasoning, and memory (Cahan et al., 2008; Ceci, 1991; Kosmidis et al., 2006, 2004; Ostrosky et al., 1998).

Considering the variability of education across cultures and the powerful influence of education on neuropsychological performance, this chapter discusses its possible role in the explanation of cross-cultural cognitive performance differences.

The influence of education on different cognitive functions

Not every cognitive function is equally influenced by schooling (Ardila & Rosselli, 2007). Data extracted from Ostrosky et al. (1998) are plotted in Figure 4.1 showing this effect. As can be observed, the educational effect on space orientation, word recognition, and motor performance-opposite reactions subtests

DOI: 10.4324/9781003051497-5

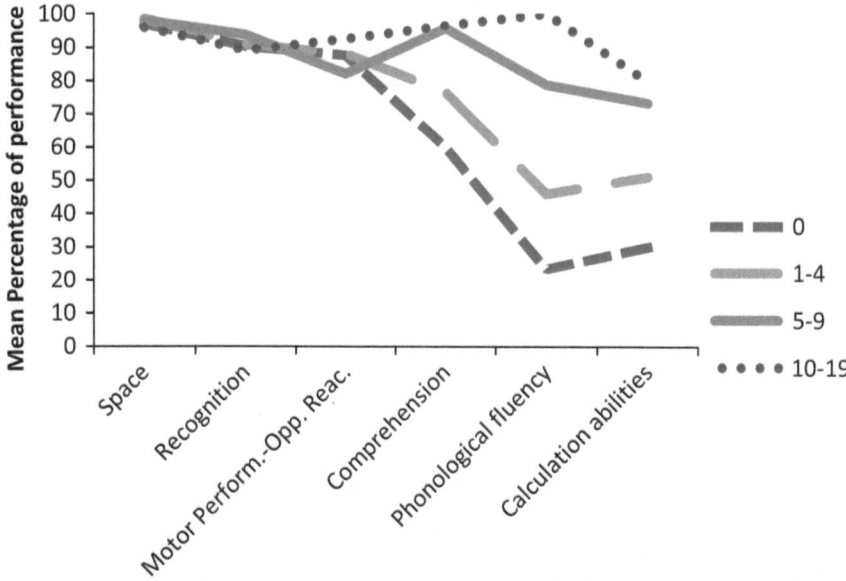

FIGURE 4.1   Mean percentage of performance by different educational groups in a sample of neuropsychological subtests. (Data adapted from Ostrosky et al., 1998).

is insignificant. But the influence of years of education is highly significant on verbal comprehension, phonological fluency, and calculation abilities subtests.

Ardila (1996) suggested that the influence of education on neuropsychological tests takes the form of a negatively accelerated curve, tending to a plateau, because of the usually low ceiling of neuropsychological tests. Indeed, data are taken from Ostrosky et al., (1998) demonstrate that effect in Figure 4.2. In this figure, it is clear that the line is steeper when it involves the lower education groups. This effect is observable in attention as well as reasoning, calculation, and verbal memory subtests.

Education or "culture"?

There are numerous examples of differences in the performance on neuropsychological tests across countries/cultures (Fernández & Marcopulos, 2008; Ostrosky-Solís & Lozano, 2006; Pérez-García et al., 2017; Salmon et al., 1989).

There are of course many differences between cultures/countries that could potentially explain performance differences and it is often difficult to determine which variables, or combination of variables, explain differences in test performance (see Chapter 2). Nevertheless, it is clear that one of the strongest explanatory factors for cross-cultural differences in performance on neuropsychological tests is education. Considering the strong influence of education on test performance and differences in education between some countries, differences in neuropsychological

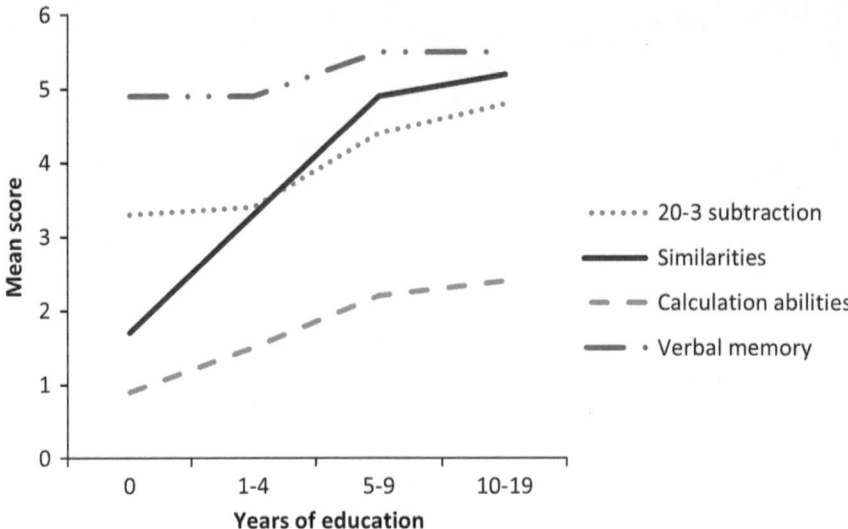

FIGURE 4.2 Mean scores by different educational groups in a sample of neuropsychological subtests (Data adapted from Ostrosky et al., 1998).

test performance across countries might be largely explained by quantitative and qualitative differences in education. Evidence supporting this hypothesis stems from the geographic distribution of IQ (a proxy for neuropsychological test performance, see Diaz-Asper et al., 2004) which matches the distribution of literacy, i.e., the higher literacy rates the higher IQ (see Figures 4.3 and 4.4[1,2]). Moreover, IQ is positively correlated with educational attainment across countries (Lynn & Mikk, 2007). Therefore, in those countries where literacy is higher, IQ is higher; in those countries where IQ is higher neuropsychological test performance is expected to be higher. There is evidence of correlations between IQ and performance on a wide range of other neuropsychological tests across-cultures, including in adult and child samples from the U.S.A, United Kingdom (Warner et al., 1987), Norway (Mohn et al., 2014), Spain (Rodriguez-Toscano et al., 2020), Finland, Egypt (Elsheikh et al., 2016), Canada (Sherman et al., 1995), and Brazil (Roque et al., 2011).

Moreover, evidence shows that significant differences in the performance of samples from two or more different countries, culturally alike, might be explained by differences in education, rather than in other cultural variables. Ramírez et al. (2005) found that there were significant differences in the performance on a verbal fluency test between Spanish-speakers from Mexico, Argentina, and Spain. However, these differences disappeared when education was controlled.

There are several studies showing that the influence of education on neuropsychological testing is as powerful as, or even stronger than, that of other aspects of culture. In a small study, Ostrosky-Solís et al. (2004) compared four groups in the Mexican Republic: (A) illiterate indigenous (Maya); (B) illiterate nonindigenous (urban dwellers); (C) indigenous with 1–4 years of education; (D)

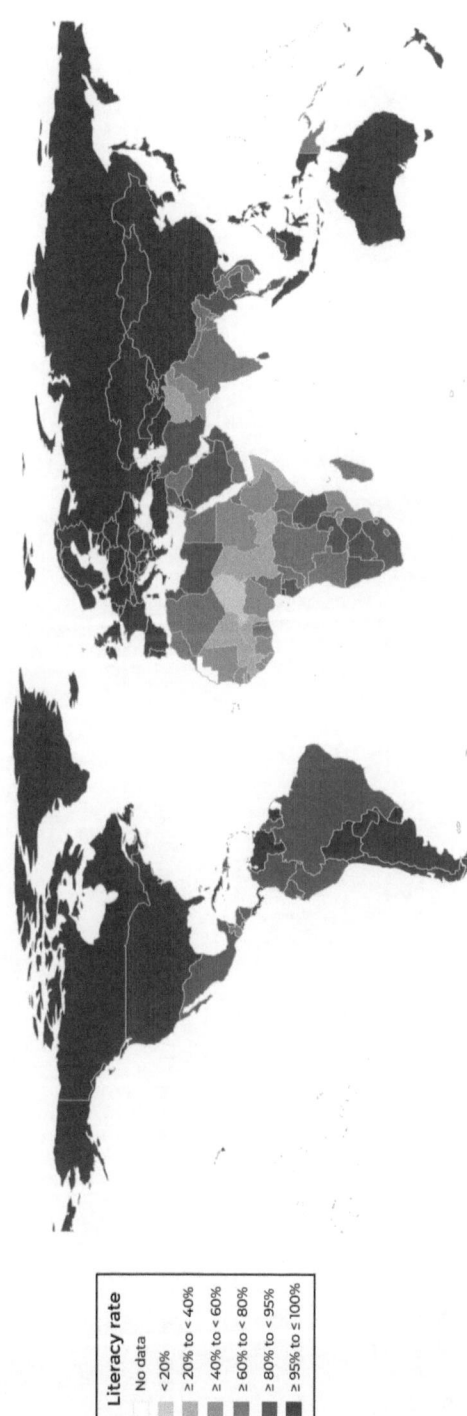

FIGURE 4.3 Adult literacy rates in the world (Roser & Ortiz-Ospina, 2016).

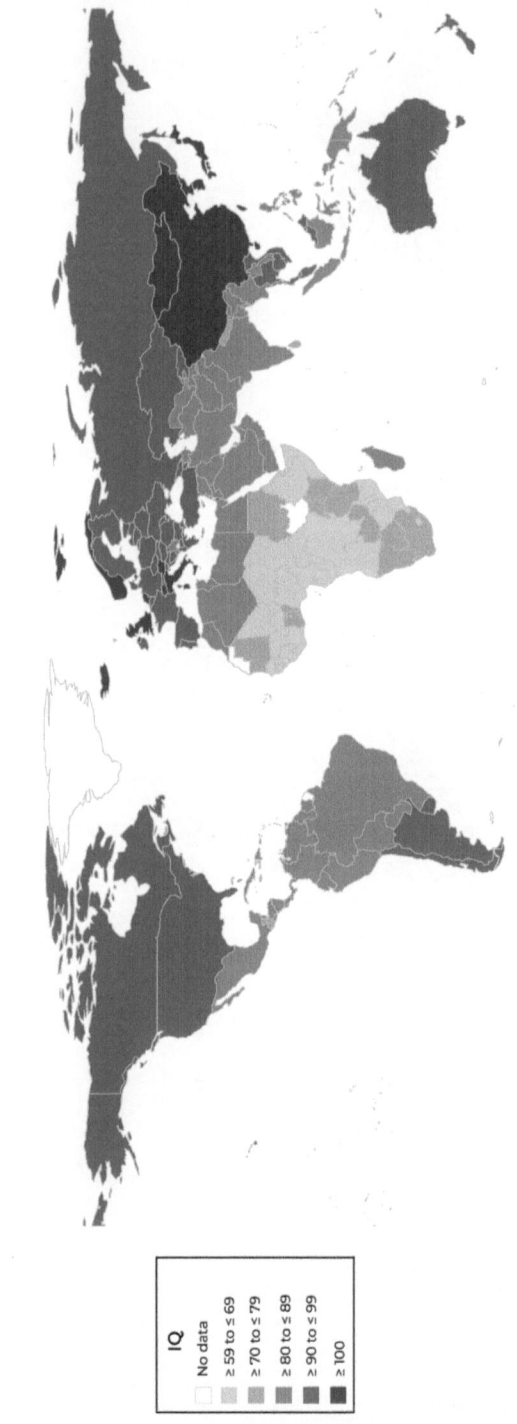

FIGURE 4.4 Mean IQ by country (Lynn & Vanhanen, 2006).

nonindigenous with 1–4 years of education. They administered the NEUROPSI neuropsychological battery to the four groups. Although they found a four-point difference in the total score between the indigenous with 1–4 years of education and the nonindigenous with 1–4 years of education (which was significant), they did not find a significant difference between illiterate indigenous and illiterate nonindigenous participants in the total score. They found significant differences between the cultural groups with 1–4 years of education (indigenous and non-indigenous) in only two subtests out of 20. However, when different education groups were compared within every culture the differences were far larger than when different cultural groups were compared: educated indigenous out-performed the illiterate indigenous by 20 points on the total score, and educated nonindigenous obtained 22 more points than the illiterate nonindigenous parti-cipants. Thus, education appeared to show a much larger effect on neu-ropsychological tests than cultural differences.

Gilleard and Gilleard (1989) compared the performance of their samples of Turkish participants with the performance of American and British participants on the Wechsler Memory Scale. They did not find significant differences across the different samples when they were education matched. However, they did find significant differences between different education groups among the Turkish participants.

Oberg and Ramírez (2006) showed a remarkable similarity in the performance of four linguistic groups (Danish, Spanish, English, and Hebrew) on a phonolo-gical fluency task. These groups belong to seemingly different cultures. Nonetheless, it is well established that education exerts a significant influence on this task (Lezak et al., 2012), suggesting that where cross-cultural differences ap-pear to be present, the underlying cause of the performance difference is likely to be related to differences in quantity or quality of education.

Harris et al. (2007) compared the performance of three different linguistic/cultural samples (English, English-Spanish bilinguals, dominant Spanish) on two attention tests. They did not find significant differences between groups. Nevertheless, they found statistically significant differences in the performance of higher and lower educated individuals within the dominant Spanish group, which varied much more widely in education than the other two groups.

Byrd et al. (2005) made the important point that a multitude of studies have compared Caucasian groups with various ethnic minority groups, finding differ-ences. But this "Caucasian-comparative" approach is limited in its ability "to identify the source of ethnicity-related variance on neuropsychological tests" (p. 1057). They argue that "within-group" studies are more useful in this respect as they allow for comparison between sub-groups that will differ in fewer char-acteristics. They compared Caribbean-born and U.S.-born African American el-derly participants on a neuropsychological battery. Interestingly they found that there was no overall difference between the groups, but reading level (which they considered to be a measure of the quality of education) was a strong predictor of performance.

Greenfield (1997) suggests that "a major (probably the major) factor that makes a culture more or less different from the cultural conventions surrounding ability testing is the degree of formal education possessed by the participants" (p. 1119); and she adds "where high levels of formal education are equated across cultures, I would predict that the values, epistemology, and communicative conventions of the participants would be extremely similar to each other" (p. 1119).

All things considered, these results might be analyzed from a different point of view. These and other studies are comparing education versus a group of variables such as culture, nationality, or linguistic group. However, the definition of culture is unspecified and varies from one study to another. In many cases, different culture means different nationality or geographical region. That is the case for the studies comparing "Westerners" and "Easterners". Some questions should be asked in this case: where is the limit between the West and the East? Are all Westerners (or Easterners) part of the same culture? For example, can we assume that Indian and Japanese cultures are so alike to be included in the same cultural group? Where do the African cultures fit? Are they Western or Eastern cultures? Can we consider Egypt and South Africa part of the same "African" culture? Moreover, this approach assumes that these compared cultures have very clearly defined borders and are distinguishable from each other. The acculturation phenomenon is often neglected despite its strong influence (see Chapters 2 and 12). This might be, for example, an explanation for the findings in the Ostrosky-Solís et al. study. It is likely that their indigenous population is highly integrated into the Western culture. In fact, they state that new generations speak more Spanish than Mayan. Therefore, rather than education versus culture, in future studies, the analysis should be focused on education versus other operationally defined cultural variables (see Chapter 2).

To sum up, although a large number of cultural variables might be causing the differences found in the neuropsychological test performance between cultural groups it seems that quantity and quality of education might explain a large amount of these differences. This variable, as well as age, should always be properly examined/controlled before reaching conclusions about cultural differences.

Quantity versus quality of education

Researchers have frequently used the number of years of education to quantify the level of education. This method assumes that the more years a participant has attended school the more education that person has received. Although this method is useful and seems to be a reasonable proxy of the educational level of a participant, it has been increasingly questioned in the last two decades. The issue is that this approach assumes that the quality of education is similar between two participants having the same number of years of education. However, that assumption is usually not possible to demonstrate, and in many cases, it is most likely false. It is a fact that the quality of education varies across different countries, regions, states, and even schools within the same city. Several studies have shown that some underprivileged social groups receive lower quality education compared to affluent social groups

(Chin et al., 2012; Manly et al., 2002; Shuttleworth-Edwards et al., 2004). Several factors such as contents of the syllabus, teaching style, funding, teacher quality, student/teacher ratios, and access to special facilities or length of school year vary across schools and have an impact on educational attainment. Thus, two participants having the same number of years of education may have a quite different quality of education. This introduces a nuance factor into research studies trying to determine the influence of education on neuropsychological testing performance. Indeed, it is also relevant for studies examining other aspects of culture where the intention is to control for education – it may be that years of education are controlled for, but this may not effectively control for quality of education.

Furthermore, alternative approaches to estimate educational level have produced additional evidence demonstrating flaws with the standard method. For example, Lam et al. (2013) used an index that they named Adjusted Years of Education in which they take into consideration the presence of alternative choices to mainstream education. In this index, education is divided into 12 categories, and it is derived by summing the adjusted years of education at each stage. In their study, they found that this index correlated more highly than total years of education with scores from neuropsychological tests.

A different approach has been attempted to overcome these problems. Some authors have used reading skills as a proxy for the quality of education. Because reading is a skill generally acquired at school and is directly associated with the quality of education received, it is considered a more accurate indicator of educational level (Sayegh et al., 2014).

Using reading skills as a proxy for quality of education has been successful in explaining differences in the performance of different groups that were matched in the number of years of education. Pawlowski et al. (2012) showed that reading and writing frequency are highly associated with neuropsychological performance. They found significant differences in performance on neuropsychological measures of attention, memory, arithmetic abilities, language, constructional praxis, problem solving, and verbal fluency tasks between low and high frequency of reading and writing groups. This difference persisted between participants with the same number of years of education but different reading and writing habits. Manly et al. (2002) showed that differences in the performance on neuropsychological tests between African Americans and non-Hispanic Whites were greatly reduced and racial differences on almost all tests became non-significant when their performances were adjusted by their reading scores. In addition, Schafer Johnson et al. (2006) found that reading ability correlated significantly with the performance on executive functioning tests. In a multiple regression analysis, they even found that when the reading ability was included in the analysis, the correlation with years of education became non-significant in some cases, suggesting that reading ability was mediating the relationship between education and test performance. Moreover, they found that 73.19% of the participants read at a grade level below their stated educational attainment, clearly indicating that the number of years of education does not always correlate with the expected level of cognitive development. The latter finding is in agreement with other studies (Baker et al., 1996; O'Bryant et al., 2007).

In a recent study, Fernández and Jáuregui (2021) used reading fluency as a measure of educational level. Reading fluency is "the oral translation of text with speed and accuracy" (Fuchs et al., 2001, p. 239). They found that in a sample of healthy controls reading fluency correlated higher than number of years of education with a neuropsychological screening test ($r = 0.53$ versus $r = 0.38$). The authors advocate this approach as it represents some advantages over reading irregular words: it is shorter and can be applied in transparent and opaque languages as well.

In summary, there is a growing consensus that quality of education needs to be considered when the influence of demographic variables on neuropsychological performance is examined. Reading skills, with different methods, is currently the most widely accepted index as a proxy for quality of education. However, only a small percentage of studies report this index. Measuring reading skills is not a difficult task and it seems appropriate to include it as a routine procedure within studies testing the influence of educational level on neuropsychological performance to better understand this relationship.

Illiteracy and culture

Throughout this chapter there has been an extensive discussion of the influence of schooling on neuropsychological performance across cultures. There are excellent articles discussing the performance of people who are illiterate on neuropsychological tests (Ardila et al., 2010), but what about the cognitive functioning of illiterate people across cultures? Do they perform differently on neuropsychological tests? Are illiterate people cognitively different across cultures?

As stated above, literacy has a significant impact on neuropsychological performance by changing the dynamics of brain functioning. Schooling seemingly tends to smooth the differences on neuropsychological performance between cultures; therefore, it is interesting to test if there are differences in cognitive performance in those individuals who have not attended school. If schooling tends to equalize the performance across cultures, then those who are illiterate, unaffected by the influence of schooling, might show more clearly the impact of other cultural variables on neuropsychological performance.

Information regarding this topic is scarce. Research with illiterate people has been conducted in different countries. Nevertheless, few or no comparisons have been attempted across illiterate people from different cultures. Comparing performance on specific tests is also hindered by the lack of equivalency between different versions of the same test. Because some tests have been modified across cultures, versions are not equivalent and therefore data are not comparable. For example, in the language section of the Chinese version of the Mini-Mental State Examination (MMSE) participants were asked to say a sentence, in the Bangla version they were asked to answer a question and in the Indian version they were just asked to say something about their house (Ganguli et al., 1995; Kabir & Herlitz, 2000; Katzman et al., 1988).

It is not possible to determine if the difficulty of these activities is comparable across versions without empirical data to test it. Some adaptations might be even tapping on a different cognitive domain. For example, in the Bangla version of the MMSE participants were asked to calculate the rickshaw (a transport) fare rather than subtracting 7s backward, which raises doubts about the cognitive process involved.

Table 4.1 presents the results of an analysis comparing the performance of different samples of illiterate people on the animal fluency test (these data were

TABLE 4.1 Comparison of the performance between illiterate samples across cultures

Place of Origin of the Sample	n	Mean Ages	Means	SD's	p	References
Greece-U.S.A.	19–43	72–74.8	11.8–11.5	3.2–3.1	$p = 0.73$	Kosmidis et al. (2004)-Manly et al. (1999)
Greece-Mexico	19–50	72–71.2	11.8–13.1	3.2–7.1	$p = 0.30$	Kosmidis et al. (2004)-Ostrosky et al. (1999)
Mexico-U.S.A.	50-43	71.2–74.8	13.1-11.5	7.1-3.1	$p = 0.15$	Kosmidis et al. (2004)-Manly et al. (1999)
Mexico-Mexico	10–50	38.2–39.6	13.5–13.7	3–4.5	$p = 0.89$	Montiel & Matute (2006)-Ostrosky et al. (1999)
Mexico-Mexico	49-7	58.8-58.4	12.7-11	5-2.5	$p = 0.36$	Ostrosky et al. (1999)-Ostrosky-Solis et al. (2004)
Mexico-Pame Indians (Mexico)	49-6	58.8-53.6	12.7-11.8	5-4	$p = 0.67$	Ostrosky et al. (1999)-Ostrosky-Solis et al. (2004)
Mexico-Portugal	49-23F	58.8–62	12.7-10.8	5-3.7	$p = 0.09$	Ostrosky et al. (1999)-Reis and Castro-Caldas (1997)
Maya Indians (Mexico)-Pame Indians (Mexico)	7-6	58.4-53.6	11–11.8	2.5–4	$p = 0.67$	Ostrosky and colleagues (2004)-Ostrosky-Solis et al. (2004)
Maya Indians (Mexico)-Portugal	7–23F	58.4–62	11-10.8	2.5–3.7	$p = 0.90$	Ostrosky and colleagues (2004)-Reis and Castro-Caldas (1997)
Pame Indians (Mexico)-Portugal	6–23F	53.6–62	11.8-10.8	4-3.7	$p = 0.57$	Ostrosky-Solis et al. (2004)-Reis and Castro-Caldas (1997)

compiled by Ardila et al., 2010). Comparisons were made on the basis of groups with similar mean ages. Overall, there are no significant differences in any of these comparisons. These data seem to indicate that performance on this task is not affected by culture, although should be considered preliminary because of methodological issues. It is remarkable that some of these analyses compare Mexican Indians with the urban population.

Conclusions

Education is one of the most powerful variables influencing performance on neuropsychological tests. Different studies show that many of the apparent cultural differences might be explained by differences in the quantity or quality of education. Interestingly, some preliminary data with illiterates do not show differences across cultures. Thus, it is likely that a large part of the differences found between cultures is also linked to culturally biased methods.

Finally, this chapter has highlighted the importance of using alternative methods to measure the educational level of participants.

These conclusions lead to several considerations. First is the importance of schooling in the comparison of performances between different cultural samples. This variable should be carefully controlled in order to determine its possible influence on the results. Second are the implications for test development. Most current neuropsychological tests are developed for highly educated individuals. In light of the strong influence of schooling, test developers should carefully consider the development of new tests that are appropriate for low educated and illiterate individuals, taking into account that a large proportion of world's population falls in these categories (see Chapter 10). Finally, there are clinical implications that mainly involve the interpretation of performance and especially test results when assessing low-educated people. Developing normative data for this population is necessary in order to deliver a fair and accurate assessment of their cognitive status.

Notes

1 When correlated, these data yielded a $r = 0.64$, $p < 0.05$ ($n = 189$) – a ceiling effect on the literacy variable probably prevented a higher correlation coefficient.
2 Adult literacy rate is the percentage of people aged 15 and above who can read and write with some understanding a short simple statement about their everyday life. Definitions may differ in some countries.

References

Ardila, A. (1996) Towards a cross-cultural neuropsychology. *Journal of Social and Evolutionary Systems*, *19*(3):237–248.
Ardila, A. (2007). Toward the development of a cross-linguistic naming test. *Archives of Clinical Neuropsychology*, *22*, 297–307.

Ardila, A., Bertolucci, P.H., Braga, L.W., Castro-Caldas, A., Judd, T., Kosmidis, M.H., … Rosselli, M. (2010) Illiteracy: The neuropsychology of cognition without reading. *Archives of Clinical Neuropsychology, 25*:689–712.

Ardila, A., & Rosselli, M. (2007). Illiterates and cognition: The impact of education. In B.P. Uzzell, M. Pontón, & A. Ardila (Eds.), *International handbook of cross-cultural neuropsychology.* (pp. 181–198). Mahwah, New Jersey: Lawrence Erlbaum Associates.

Baker, F.M., Johnson, J.T., Velli, S.A., & Wiley, C. (1996). Congruence between education and reading levels of older persons. *Psychiatric Services, 47*(2), 194–196. 10.1176/ps.47.2.194.

Byrd, D.A., Sanchez, D. & Manly, J.J. (2005) Neuropsychological test performance among Caribbean-born and U.S.-born African American elderly: The role of age, education and reading level. *Journal of Clinical and Experimental Neuropsychology, 27*(8), 1056–1069.

Cahan, S., Greenbaumb, C., Artman, L., Deluya, N. & Gappel-Gilon, Y. (2008) The differential effects of age and first grade schooling on the development of infralogical and logico-mathematical concrete operations. *Cognitive Development, 23*, 258–277.

Castro-Caldas, A., & Reis, A. (2000). Neurobiological substrates of illiteracy. *The Neuroscientist, 6*(6), 475–482. 10.1177/107385840000600610.

Ceci, S.J. (1991) How much does schooling influence general intelligence and its cognitive components? A reassessment of the evidence. *Developmental Psychology, 27*(5), 703–722.

Chin, A.L., Negash, S., Xie, S., Arnold, S.E., & Hamilton, R. (2012). Quality, and not just quantity, of education accounts for differences in psychometric performance between African Americans and White Non-Hispanics with Alzheimer's disease. *Journal of the International Neuropsychological Society, 18*(2), 277–285. 10.1017/s1355617711001688.

Diaz-Asper, C.M., Schretlen, D.J., & Pearlson, G.D. (2004). How well does IQ predict neuropsychological test performance in normal adults? *Journal of the International Neuropsychological Society, 10*(01). 10.1017/s1355617704101100.

Elsheikh, S., Kuusikko-Gauffin, S., Mattila, M.-L., Jussila, K., Ebeling, H., Loukusa, S., … Moilanen, I. (2016). Neuropsychological performance of Finnish and Egyptian children with autism spectrum disorder. *International Journal of Circumpolar Health, 75*(1), 29681. 10.3402/ijch.v75.29681.

Fernández, A.L. & Jáuregui, G.E. (2021). Reading fluency as a measure of the educational level. *Dementia & Neuropsychologia, 15*(3), 361–365.

Fernández, A.L. & Marcopulos, B. (2008). A comparison of normative data for the Trail Making Test from several countries: Equivalence of norms and considerations for interpretation. *Scandinavian Journal of Psychology, 49*, 239–246.

Fuchs, L.S., Fuchs, D., Hosp, M.K., & Jenkins, J.R. (2001). Oral reading fluency as an indicator of reading competence: A theoretical, empirical, and historical analysis. *Scientific Studies of Reading, 5* (3), 239–256.

Ganguli, M., Ratcliff, G., Chandra, V., Sharma, S., Gilby, J., Pandav, R., … Dekosky, S. (1995). A hindi version of the MMSE: The development of a cognitive screening instrument for a largely illiterate rural elderly population in india. *International Journal of Geriatric Psychiatry, 10*(5), 367–377. 10.1002/gps.930100505.

Gilleard, E., & Gilleard, C. (1989). A comparison of Turkish and Anglo-American normative data on the Wechsler memory scale. *Journal of Clinical Psychology, 45*(1), 114–117. 10.1002/1097-4679(198901)45:13.0.co;2-8.

Greenfield, P.M. (1997). You can't take it with you: Why ability assessments don't cross cultures. *American Psychologist, 52*(10), 1115–1124. 10.1037/0003-066X.52.10.1115.

Grossi, D., Correra, G., Calise, C., Ruscitto, M.A., Vecchione, V., Vigliardi, M.V., & Nolfe, G. (1993). Evaluation of the influence of illiteracy on neuropsychological performances by elderly persons. *Perceptual and Motor Skills*, 77, 859–866.

Harris, J.G., Wagner, B. & Munro Cullum, C. (2007) Symbol vs. digit substitution task performance in diverse cultural and linguistic groups, *The Clinical Neuropsychologist*, 21(5), 800–810, DOI: 10.1080/13854040600801019.

Kabir, Z.N., & Herlitz, A. (2000). The Bangla Adaptation of Mini-mental State Examination (BAMSE): An instrument to assess cognitive function in illiterate and literate individuals. *International Journal of Geriatric Psychiatry*, 15(5), 441–450. 10.1002/(sici)1099-1166(200005)15:53.0.co;2-o.

Katzman, R., Zhang, M., Ouang-Ya-Qu, Wang, Z., Liu, W.T., Yu, E., ... Grant, I. (1988). A Chinese version of the mini-mental state examination; Impact of illiteracy in a Shanghai dementia survey. *Journal of Clinical Epidemiology*, 41(10), 971–978. 10.1016/0895-4356(88)90034-0.

Kosmidis, M.H., Tsapkini, K., & Folia, V. (2006). Lexical processing in illiteracy: Effect of literacy or education? *Cortex*, 42, 1021–1027.

Kosmidis, M.H., Tsapkini, K., Folia, V., Vlahou, C.H., & Kiosseoglou, G. (2004). Semantic and phonological processing in illiteracy. *Journal of the International Neuropsychological Society*, 10, 818–827.

Lam, M., Eng, G.K., Rapisarda, A., Subramaniam, M., Kraus, M., Keefe, R.S.E., & Collinson, S.L. (2013). Formulation of the age–education index: Measuring age and education effects in neuropsychological performance. *Psychological Assessment*, 25(1), 61–70. 10.1037/a0030548.

Lezak, M.D., Howieson, D.B., Bigler, E.D. & Tranel, D. (2012) *Neuropsychological Assessment*. Fifth edition. New York: Oxford.

Lynn, R., & Mikk, J. (2007). National differences in intelligence and educational attainment. *Intelligence*, 35(2), 115–121. 10.1016/j.intell.2006.06.001.

Lynn, R., & Vanhanen, T. (2006). *Iq and global inequality*. Washington Summit Publishers.

Manly, J.J., Jacobs, D.M., Sano, M., Bell, K., Merchant, C.A., Small, S., et al. (1999). Effect of literacy on neuropsychological test performance in nondemented, education-matched elders. *Journal of the International Neuropsychological Society*, 5, 191–202.

Manly, J.J., Jacobs, D.M., Touradji, P., Small, S.A., & Stern, Y. (2002). Reading level attenuates differences in neuropsychological test performance between African American and White elders. *Journal of the International Neuropsychological Society*, 8(3), 341–348. 10.1017/s1355617702813157.

Marcopulos, B.A., McLain, C.A., & Giuliano, A.J. (1997). Cognitive impairment or inadequate norms? A study of healthy, rural, older adults with limited education. *The Clinical Neuropsychologist*, 11, 111–131.

Mohn, C., Sundet, K., & Rund, B.R. (2014). The relationship between IQ and performance on the MATRICS consensus cognitive battery. *Schizophrenia Research: Cognition*, 1(2), 96–100. 10.1016/j.scog.2014.06.003.

Montiel, T., & Matute, E. (2006). La relación entre alfabetización y la escolarización con el desempeño en tareas verbales. In E. Matute (Ed.), *Lectura y diversidad cultural*. Guadalajara, Mexico: Universidad de Guadalajara.

Nell, V. (2000). *Cross-cultural neuropsychological assessment: Theory and practice*. Mahwah, NJ: Lawrence Erlbaum Associates.

Oberg, G., & Ramírez, M. (2006). Cross-linguistic meta-analysis of phonological fluency: Normal performance across cultures. *International Journal of Psychology*, 41(5), 342–347. 10.1080/00207590500345872.

O'Bryant, S., Lucas, J., Willis, F., Smith, G., Graff-Radford, N., & Ivnik, R. (2007). Discrepancies between self-reported years of education and estimated reading level among elderly community-dwelling African-Americans: Analysis of the MOAANS data. *Archives of Clinical Neuropsychology, 22*(3), 327–332. 10.1016/j.acn.2007.01.007.

Ostrosky, F., Ardila, A., & Rosselli, M. (1999). NEUROPSI: A brief neuropsychological test battery in Spanish. *Journal of the International Neuropsychological Society, 5,* 413–433.

Ostrosky, F., Ardila, A., Rosselli, M., López-Arango, G., & Uriel-Mendoza, V. (1998). Neuropsychological test performance in illiterates. *Archives of Clinical Neuropsychology, 13,* 645–660.

Ostrosky-Solís, F., & Lozano, A. (2006). Digit Span: Effect of education and culture. *International Journal of Psychology, 41*(5), 333–341. 10.1080/00207590500345724.

Ostrosky-Solis, F., Ramírez, M. & Ardila, A. (2004). Effects of culture and education on neuropsychological testing: A preliminary study with indigenous and nonindigenous population. *Applied Neuropsychology, 11*(4), 186–193, DOI: 10.1207/s15324826an1104_3.

Ostrosky-Solís, F., Ramírez, M., Lozano, A., Picasso, H., & Vélez, A. (2004). Culture or education? Neuropsychological test performance of a Maya indigenous population. *International Journal of Psychology, 39* (1), 36–46.

Pawlowski, J., Remor, E., Pimenta Parente, M.A., de Salles, J.F., Fonseca, R.P., & Bandeira, D.R. (2012). The influence of reading and writing habits associated with education on the neuropsychological performance of Brazilian adults. *Reading and Writing, 25*(9), 2275–2289. 10.1007/s11145-012-9357-8.

Pérez-García, M., Luna, J.D.D., Torres-Espínola, F.J., Martínez-Zaldívar, C., Anjos, T., Graaff, J.S.-D., … Campoy, C. (2017). Cultural effects on neurodevelopmental testing in children from six European countries: An analysis of NUTRIMENTHE Global Database. *British Journal of Nutrition, 122*(s1). 10.1017/s0007114517000824.

Ramírez, M., Ostrosky-Solís, F., Fernández, A., Ardila-Ardila, A. (2005). Fluidez verbal semántica en hispanohablantes: Un análisis comparativo (Semantic verbal fluency in Spanish-speaking population: A comparative analysis). *Revista de Neurología, 41* (8): 463–468.

Reis, A., & Castro-Caldas, A. (1997). Illiteracy: A bias for cognitive development. *Journal of the International Neuropsychological Society, 3,* 444–450.

Rodriguez-Toscano, E., López, G., Mayoral, M., Lewis, S., Lees, J., Drake, R., Arango, C., & Rapado-Castro, M. (2020). A longitudinal comparison of two neurocognitive test batteries in patients with schizophrenia and healthy volunteers: Time effects on neuropsychological performance and their relation to functional outcome. *Schizophrenia research, 216,* 347–356. 10.1016/j.schres.2019.11.018.

Roque, D.T., Teixeira, R.A.A., Zachi, E.C. & Ventura, D.F. (2011). The use of the Cambridge Neuropsychological Test Automated Battery (CANTAB) in neuropsychological assessment: Application in Brazilian research with control children and adults with neurological disorders. *Psychology & Neuroscience, 4*(2), 255–265. 10.3922/j.psns.2011.2.011.

Roser, M. & Ortiz-Ospina, E. (2016). Literacy. Published online at OurWorldInData.org. Retrieved from: https://ourworldindata.org/literacy [Online Resource]

Salmon, D.P., Riekkinen, P., Katzman, R., Zhang, M., Jin, H., & Yu, E. (1989). Cross-cultural studies of dementia: A comparison of Mini-Mental State examination performance in Finland and China. *Archives of Neurology, 46,* 769–772.

Sayegh, P., Arentoft, A., Thaler, N.S., Dean, A.C., & Thames, A.D. (2014). Quality of education predicts performance on the wide range achievement test-4th edition word reading subtest. *Archives of Clinical Neuropsychology, 29* (8), 731.

Schafer Johnson, A., Flicker, L.J., & Lichtenberg, P.A. (2006). Reading ability mediates the relationship between education and executive function tasks. *Journal of the International Neuropsychological Society, 12*(1), 64–71. 10.1017/s1355617706060073.

Sherman, E.M.S., Strauss, E., Spellacy, F., & Hunter, M. (1995). Construct validity of WAIS – R factors: Neuropsychological test correlates in adults referred for evaluation of possible head injury. *Psychological Assessment, 7*(4), 440–444. 10.1037/1040-3590.7.4.440.

Shuttleworth-Edwards, A.B., Kemp, R.D., Rust, A.L., Muirhead, J.G., Hartman, N.P., & Radloff, S.E. (2004). Cross-cultural effects on IQ test performance: A review and preliminary normative indications on WAIS-III test performance. *Journal of Clinical and Experimental Neuropsychology, 26*(7), 903–920. 10.1080/13803390490510824.

Warner, M.H., Ernst, J., Townes, B.D., Peel, J. & Preston, M. (1987) Relationships between IQ and neuropsychological measures in neuropsychiatric populations: Within-laboratory and cross-cultural replications using WAIS and WAIS-R. *Journal of Clinical and Experimental Neuropsychology, 9*(5), 545–562, DOI: 10.1080/01688638708410768.

PART II

The influence of culture on cognitive functioning

5

CONSIDERING CULTURE IN THE NEUROPSYCHOLOGICAL ASSESSMENT OF ATTENTION AND PERCEPTION

Matthew J. Russell, Hajin Lee, Karen K. Leung, and Takahiko Masuda

Introduction

Over 30 years ago, cultural psychology – an interdisciplinary field that integrates psychology, anthropology, linguistics, philosophy, and neuroscience – emerged (Bruner, 1990; Geertz, 1973). Since the advent of cultural psychology, the idea that culture and the psyche mutually constitute each other has been frequently visited (Kitayama & Cohen, 2019). Here, "culture" is defined as a pattern of beliefs, values, and practices that constitute one's environment. A plethora of empirical studies has demonstrated substantial cultural variations in psychological processes, including attention and perception (e.g., Masuda et al., 2019). We define *attention* as a process of focusing cognitive resources on aspects of the environment, and *perception* as the organization and interpretation of sensory information. Our definition of perception includes social perception, which is the organization and interpretation of social events, and is based on the premise that our social, cultural experiences fundamentally affect how we attend to and perceive the world.

This chapter presents an overview of research on culture, perception, and attention, which provides evidence that culture deeply affects basic psychological processes that relate to neuropsychological testing. This chapter does not introduce all cultural differences that affect attention and perception, but attempts to provide select examples of findings that might inform neuropsychological testing. First, we provide an overview of the frameworks of social orientation and thinking styles, and their relationship to attention and perception. Second, we provide specific examples of cultural differences that might affect neuropsychological testing. Third, we introduce research on how culture affects how individuals perceive mental illness symptoms. Finally, we introduce research showing that culture is nuanced.

DOI: 10.4324/9781003051497-7

Our objective for this chapter is to show practitioners how cultural differences in attention and perception can affect neuropsychological testing in a plethora of domains.

Social orientation, attention, and perception

Much research in cultural psychology has followed the overarching theoretical framework of *social orientation* (e.g., Markus & Kitayama, 1991; Varnum et al., 2010). In this framework, researchers have demonstrated substantial variations in cognition, attention, and perception. Those who live in a culture with a dominant *independent social orientation* (e.g., Canada or the United States) tend to view themselves as separate from their social others, and hold cognitive styles that emphasize self-direction, autonomy, and self-expression. In contrast, those who live in a culture with a dominant *interdependent social orientation* (e.g., Japan, China, and India) tend to view themselves as socially interrelated and connected to significant relationships, and hold cognitive styles that emphasize harmony, relatedness, and connection.

Researchers have provided a plethora of evidence that recurrent exposure to each social orientation influences people's thinking styles, directly influencing their cognition, attention, and perception (Varnum et al., 2010). Under the framework of *thinking styles*, Nisbett and colleagues discuss that the *analytic thinking style* is characterized by an emphasis on taxonomic and rule-based categorization, dispositional orientation in causal attribution and social inferences, formal logic in reasoning, and selective attention to foreground events at the expense of their context. In contrast, the *holistic thinking style* is characterized by thematic and family-resemblance-based categorization of objects, situational orientation in causal attribution and social inference, dialectical logic in reasoning, and sensitivity to contextual information when attending to an event (Masuda et al., 2019; Nisbett, 2003).

Thinking styles and cultural variations in attention and perception

Empirical studies have demonstrated that those with interdependent social orientation/holistic thinking styles are more likely to hold context-sensitive, holistic patterns of attention. In contrast, those with independent social orientation/analytic thinking styles endorse object-oriented, analytic patterns of attention. For example, when asked to describe short animation vignettes, European Americans refer mostly to focal objects, whereas Japanese refer more equally to both focal objects and contextual information. Such culturally variant patterns correspond to the duration of eye-fixations to stimuli, with Japanese showing more spread attention than European Americans (e.g., Chua et al., 2005). Recent cultural neuroscientific studies have further examined to what extent cultural differences in attention and perception relate to the activation of neural responses. Using event-related potential (ERP) and functional magnetic resonance imaging (fMRI) research, much evidence has demonstrated that cultural variations in neural re-sponses underlie observed cultural differences (e.g., Goto et al., 2010; Hedden et al., 2008; Masuda et al., 2014).

In addition, cultural differences in attention and perception have been observed in socio-emotional tasks, such as those involved in face recognition (e.g., Matsumoto et al., 2010; Miyamoto et al., 2011) and emotion recognition (e.g., Masuda et al., 2008, 2012).

Section conclusion

In summary, this section provides evidence that those with a more interdependent social orientation/holistic thinking style tend to have more spread attention and perception that includes context than those with independent social orientation/ analytic thinking styles (e.g., Markus & Kitayama, 1991; Nisbett, 2003). These cultural differences are relevant to neuropsychological testing. For example, social orientation and thinking styles have been related to cultural differences in attention and perception in: memory, face and emotion attention and perception, and vi-suospatial perception (Masuda et al., 2008, 2012; Nisbett & Masuda, 2003). In addition, as cultural differences affect socio-emotional processes, they have im-plications for socio-emotional related neuropsychological testing (i.e., testing for depression). Furthermore, cultural differences in attention and perception may impact how collateral history (i.e., reports by families on the person undergoing testing) and comprehensive neuropsychological testing reports are attended to and perceived. For example, a holistic practitioner might take into account more context than an analytic practitioner. Finally, it is important to note that not all cultural differences in attention and perception are related to the introduced fra-meworks of social orientation/thinking styles. As such, we provide some examples of other cultural differences below.

How culture affects attention and perception

In this section, we provide in-depth examples of how culture has been found to affect attention and perception in domains that may relate to neuropsychological testing in memory, face and emotion attention and perception, and visuospatial perceptual tasks. The examples below are just a selected few, with other papers describing additional examples in further depth (e.g., Masuda et al., 2019) and other cultural differences besides social orientation/thinking styles that may drive variations in attention and perception (e.g., Frey et al., 2021; White, 1982; Yoo & Skovholt, 2001).

Memory

Masuda and Nisbett (2001) tested how cultural differences in attention and per-ception influence memory. East Asian and North American participants were asked to evaluate how much they liked foreground animals placed in background wildernesses, and then later, to engage in an incidental memory test. The memory test involved judging if they had previously seen animals in animal-wilderness pairs that were all possible combinations of previously or newly presented animals and wildernesses (e.g., a previously presented animal on a previously presented wilderness or a previously presented animal on a new wilderness). They found that while both cultures performed well when recognizing previously presented animals and previously presented wilderness together, both groups' accuracy decreased for incongruent images with either a newly presented animal or wilderness. East Asians did worse than North Americans with incongruent images, which they related to the social orientation/thinking style dichotomy. Expanding on this research, Chua et al. (2005) used an eye tracker in the task and found that East Asians frequently alternated their attention between focal objects and the backgrounds, whereas North Americans selectively allocated their attention to focal objects. Masuda et al. (2014) further investigated neural processing during the task. They found that East Asians (and not North Americans) who processed backgrounds more in incongruent wilderness images (suggesting they perceived the wilderness as a novel), showed a drop in memory performance. This provided evidence that those who attended to and perceived a holistic link between foreground objects and context showed interference in incongruent memory judgments and had poorer memory performance. These studies support that memory is more affected by contextual information among those with interdependent/holistic thinking styles than those with independent/analytic thinking styles.

Social orientation has also been related to what type of memory is remembered. For example, Cross et al. (2002) found that individuals with greater interdependence were more likely to remember statements about a fictional person that contained relational information; however, interdependence was not related to the memory of non-relational information.

These findings suggest that social orientation/thinking styles relate to differences in how individuals' memory is affected by contextual and relational information, which may affect neuropsychological memory tests.

Faces and emotions

Masuda et al. (2008) examined North American and East Asian attention and perception of emotions with face lineups. Participants were presented with lineups of five people and were asked to judge the center person's emotion. The lineups were either: *congruent*, in which both the center person and background figures showed similar emotions (e.g., happy surrounded by happy), and *incongruent*, in which the center person's emotion was different from that of the background persons (e.g., happy surrounded by sad). East Asians perceived more effect on the center person's emotions from background persons' emotions than North Americans. As a follow-up, Masuda et al. (2012) used an eye tracker in the task and found that East Asians more frequently alternated their attention between focal faces and the background faces than North Americans. Expanding on this research, Russell et al. (2015) investigated North American and East Asian neural patterns (i.e., event related potentials [ERP]) during the task. They found that only East Asians showed ERP brain waves related to the perception of incongruent emotions as meaningful. Furthermore, they found that individuals' social orientation beliefs related to these neural processing patterns.

In addition, cultural differences have been related to face attention and perception (Miyamoto et al., 2011). In this research, East Asians tended to hold a configural-oriented mode of attention, where they viewed the face as a whole and were sensitive to the relationship among facial parts. On the other hand, European Americans tended to hold a feature-oriented mode of attention where they attended to facial features.

These findings are important as they suggest that cultural differences may also affect face and emotion processing, both of which may be tested for in neuropsychological testing for various diagnoses, such as traumatic brain injury or stroke.

Visuospatial perception

Optical illusions and pictorial depictions have elucidated how a culture's surroundings and way of living can influence perceptual processes (Deregowski, 1989). Studies with the Ebbinghaus illusion have found cultural differences (e.g., Doherty et al., 2008). In the Ebbinghaus illusion, two equal-sized central circles are perceived as different sizes depending on whether the surrounding circles are larger or smaller. Interdependent cultures often focus on contextual features, and thus, are more susceptible to the Ebbinghaus illusion than independent cultures (Doherty et al., 2008). However, with the Himba, a seminomadic society of cattle herders, researchers observed that Himba participants were actually less susceptible than Western participants; this analytic processing style may reflect social practices in Himba culture where members precisely identify their cattle despite subtle differences in markings (de Fockert et al., 2007). These cultural differences have relevance as the illusion has been used to characterize disorganized schizophrenia (King et al., 2017).

Next, the Muller-Lyer illusion, where the lengths of two equal vertical lines were perceived as shorter or longer depending on whether the terminal arrowheads were pointed inward or outward, informed the *carpentered world* and ecological hypotheses (Deregowski, 1989). Specifically, this hypothesis states that hunter-gatherer, rural, and highly urbanized groups cultures live in environments characterized by straight lines and right angles (e.g., rooms and furniture in urban societies), and are more susceptible to this illusion (Segall et al., 1966). In contrast, cultural groups living in open terrains were susceptible to other optical effects, such as the horizontal-vertical illusion (Deregowski, 1989). The Muller-Lyer illusion is relevant to testing as it has been used to discriminate autism and visuospatial neglect (Chouinard et al., 2013; Daini et al., 2002).

Finally, although seemingly simple, cube-copying discrimination tasks have been used to assess visuospatial perception functioning in dementia (Mosimann et al., 2004).

The cube-copying test is frequently used and contains an optical illusion where the cube could be simultaneously facing either toward the lower-left or upper-right. Differences between cultures and familiarity with pictorial renderings of three-dimensional objects and optical illusions influence performance. For example, research has demonstrated that cultures with less experience with cube perception tend to score poorly despite intact cognition (e.g., Ardila & Moreno, 2001).

Together these findings suggest the importance of considering culture in visuospatial test design.

Culture, depression, and anxiety

As one example of how culture may affect socio-emotional related neuropsychological testing, research has found that culture influences how people socially perceive symptoms of depression and anxiety. These differences have been explained by various cultural mechanisms. We focus on East Asian populations as they have been identified as a vulnerable group that is at increased risk for experiencing adverse mental health outcomes (Islam et al., 2014; Kirmayer et al., 2007).

Cultural differences in symptom perception

People from Western countries tend to perceive psychological symptoms of distress (i.e., psychologization), whereas people from East Asian countries tend to perceive somatic symptoms (i.e., somatization, a physical manifestation of psychological distress; Ryder et al., 2008). For example, Chinese depressed patients report more somatic symptoms (e.g., changes in sleep, lack of energy, or loss of appetite) compared to their Canadian counterparts. In contrast, Canadian depressed patients tend to report more psychological symptoms (e.g., feeling depressed, lonely, or hopeless) than their Chinese counterparts. Additionally, populations in other regions of East Asia, including South Korea (Zhou et al., 2015) and Japan (Nakamura et al., 2017), tend to show this tendency of focusing more on somatic symptoms in their experience of distress. Non-clinical populations similarly describe their experience of distress distinctly (Lee, 2019; Tsai et al., 2004).

Cultural mechanisms

We introduced that Westerners perceive themselves as separate from others, holding an independent social orientation/analytic thinking style. In contrast, East Asians perceive themselves as interconnected with others, having an interdependent social orientation/holistic thinking style (e.g., Nisbett, 2003; Varnum et al., 2010). Along these cultural divides, Ryder et al. (2008) provided evidence that individuals' *externally-oriented thinking style* (a focus on external rather than internal states, which is conceptually similar to interdependence/holistic thinking) partially mediates the relationship between culture and depressive symptom presentation. In other words, somatization among Chinese depressed patients is thought to be explained by a holistic tendency to perceive both psychological states and physical sensations; while psychologization among European Canadians is explained by an analytic tendency to primarily perceive psychological states.

In addition, other cultural differences have been noted to relate to the perception of depression/anxiety symptoms. First, the stigma of mental illness has been shown to influence symptom perception. East Asian societies tend to hold greater stigma towards mental health disorders (Ryder et al., 2000), which has been associated with East Asian individuals mainly communicating physical symptoms to mental health practitioners. Second, this tendency has been related to cultural differences in social support-seeking strategies. Choi et al. (2016) demonstrated that Koreans tend to perceive the use of somatic words as more effective in seeking social support compared to Americans. Furthermore, somatization patterns have been associated with a delay in help-seeking and the underuse of mental health services among East Asians (White, 1982; Yoo & Skovholt, 2001). These overarching cultural differences in social perception of stigma and appropriateness of seeking mental health support may influence whether a person agrees to neuropsychological testing and affect subsequent treatment-seeking.

Section conclusion

To date, prior cultural-clinical literature has mainly used self-report questionnaires and interviews to assess individuals' psychological distress. Considering that neuropsychological tests (such as Beck's Depression Inventory and other mental health tests) are used to test for psychological distress, it is important to consider the appropriate use of these testing tools to address cultural differences in the symptom manifestation of psychological distress, as well as cultural differences in the perception of stigma and social support use. This can help us develop and implement culturally appropriate mental health services and interventions to better support individuals of different cultural backgrounds.

Cultural nuances in attention and perception

Moving from a monolithic view of cultural differences in attention and perception, where culture is seen as equal across situations and individuals, this section introduces some recent efforts to show ways culture is nuanced. We provide a few examples of boundary conditions and individual differences of culture, where: (1) culture is affected by situational context, (2) individuals differ in culture, and (3) chronic exposure to different cultural experiences changes individuals' culture.

A few examples of cultural nuances

First, it is important to note that cultural differences in attention and perception may be influenced by small contextual factors in session. For example, research has found that cultural differences can be *primed* (presented with cues from one of their cultures to activate that culture) by language or visual cues to change attention- and perception-related patterns to prime-related cultural patterns (e.g., Fong et al., 2014; Lin & Han, 2009). For example, researchers have used social orientation priming to alter individuals' attention and perception (e.g., using *I/mine* language for independent priming; and *we/ours* language for interdependent priming). Interdependent priming resulted in increasing individuals' distraction from incongruent context (e.g., letter Es surrounded by Hs; versus congruent context [Hs surrounded by Hs]) in the flanker task (an attention and executive function task), and a bias towards global perception (more holistic interpretations of stimuli) versus local perception (Lin & Han, 2009). Other research has suggested that cultural differences are affected by differences in instruction/context (e.g., Ito et al., 2013; Russell et al., 2019; Senzaki et al., 2014). For example, when East Asians and North Americans were asked to form narratives about underwater scenes, East Asians were more likely to attend to context than North Americans; however, this difference disappeared when both cultures were just asked to view the scenes (Senzaki et al., 2014). The authors proposed that cultural differences in attention and perception may appear best when individuals are asked to make meaning from the world, which is a key aspect of our definition of culture. The implication is that language use and the framing of sessions may affect neuropsychological testing, such that considering culturally-neutral, standardized procedures, and even language use, is very important.

Second, it is important to note that culture can differ among individuals within a cultural context. Research has found that individuals' self-reported cultural beliefs explain observed neural-cultural differences in attention and perception (e.g., Goto et al., 2010; Na & Kitayama, 2011; Russell et al., 2015, 2019, 2020). Despite these findings, other research suggests that current cultural self-reports sometimes do not predict cultural differences in attention and perception behavioral patterns (e.g., Na et al., 2010). This work proposed that while culture exists on the aggregate across tasks, individual differences in behavior may cohere less to the overall, aggregate culture. On the other hand, we believe that neuropsychological testing has a stronger relationship to individual differences due to its connection to underlying basic attention and perception processes. The implication is that the application of cultural self-report scales in neuropsychological testing is unlikely to be a solution for detecting individual differences in attention and perception related to culture, although such application might be possible in the future with proper development and validation.

Finally, it is important to note that individuals' experiences can change their individual-level culture. These differences may derive from various influences, such as: (1) living in a region with different historic developments (e.g., farming versus fishing communities), (2) holding different religious backgrounds, and (3) growing up rich versus poor (Kitayama & Cohen, 2019; Varnum et al., 2010). For example, lower-income people have been shown to follow attention and perception patterns that are more similar to interdependent/holistic than those who are higher-income, who show more independent/analytic patterns (Varnum et al., 2012). As a further nuance, research suggests that individual-level culture may also change over time.

For example, people that move to different cultural contexts may change cultural patterns, through *acculturation* (i.e., the process of adjusting and adapting to a new cultural context). For example, cultural differences in emotion perception were found to be reduced for bicultural Asian Canadians and Asian International students living in Canada, as compared to Japanese students living in Japan (e.g., Masuda et al., 2012). Also, in various interdependent groups in America, acculturation has been related to the performance on various neuropsychological tests related to attention and perception (Razani et al., 2007). It is important to consider these experiential differences, as beyond ethnic categories, differences in experiences (including changes in experience due to acculturation) may affect a persons' cultural background, and through it, their attention and perception.

Section conclusion

To close, this section provides context on how culture is nuanced. The section provides arguments of the importance of using culturally neutral, standardized procedures to avoid contextual influences on behaviors in neuropsychological testing, the need to develop better individual difference self-report culture measures/batteries, and the need to consider individuals' culture due to experiential differences, moving beyond the level of simple dichotomies and ethnic categories.

Chapter summary

This chapter provides evidence of how cultural differences in attention and perception can affect neuropsychological testing in a plethora of domains. First, we provided evidence of how social orientation/thinking styles are connected to cultural differences in attention and perception. Second, we provided various examples of how culture affects attention and perception. Third, we showed how culture has been linked to depression symptom perception, with East Asians reporting more somatic symptoms compared to a Western focus on psychological symptoms. These patterns were further related to cultural differences in the social perception in stigma and social support seeking. Finally, we introduced nuances of how culture may affect attention and perception through situational context, individual differences in culture patterns, and differences in culture-related experience.

This chapter shows the importance of considering culture in attention- and perception-related neuropsychological testing. It argues that culture is complex and nuanced; however, with some thought and development, it should be possible to develop ways to better implement neuropsychological tests that take into account the effect of culture.

References

Ardila, A., & Moreno, S. (2001). Neuropsychological test performance in Aruaco Indians: An exploratory study. *Journal of the International Neuropsychological Society, 7*, 510–515.

Bruner, J. (1990). *Acts of meaning.* Cambridge, MA: Harvard University Press.

Choi, E., Chentsova-Dutton, Y., & Parrott, W. (2016). The effectiveness of somatization in communicating distress in Korean and American cultural contexts. *Frontiers in Psychology, 7,* 383.

Chouinard, P.A., Noulty, W.A., Sperandio, I., & Landry, O. (2013). Global processing during the Müller-Lyer illusion is distinctively affected by the degree of autistic traits in the typical population. *Experimental Brain Research, 230,* 219–231.

Chua, F., Boland, J., & Nisbett, R. (2005). Cultural variation in eye movements during scene perception. *Proceedings of the National Academy of Sciences of the United States of America, 102,* 12629–12633.

Cross, S.E., Morris, M.L., & Gore, J.S. (2002). Thinking about oneself and others: The relational-interdependent self-construal and social cognition. *Journal of Personality and Social Psychology, 82*(3), 399–418.

Daini, R., Angelelli, P., Antonucci, G., Cappa, S.F., & Vallar, G. (2002). Exploring the syndrome of spatial unilateral neglect through an illusion of length. *Experimental Brain Research, 144,* 224–237.

de Fockert, J., Davidoff, J., Fagot, J., Parron, C., & Goldstein, J. (2007). More accurate size contrast judgments in the Ebbinghaus Illusion by a remote culture. Journal of *Experimental Psychology: Human Perception and Performance, 33,* 738.

Deregowski, J.B. (1989). Real space and represented space: Cross-cultural perspectives. *Behavioral and Brain Sciences, 12,* 51–119.

Doherty, M.J., Tsuji, H., and Phillips, W.A. (2008). The context sensitivity of visual size perception varies across cultures. *Perception 37,* 1426–1433.

Fong, M., Goto, S., Moore, C., Zhao, T., Shudson, Z., & Lewis, R. (2014). Switching between Mii and Wii: The effects of cultural priming on the social affective N400. *Culture and Brain, 2,* 52–71.

Frey, K.S., Onyewuenyi, A.C., Hymel, S., Gill, R., & Pearson, C.R. (2021). Honor, face, and dignity norm endorsement among diverse North American adolescents: Development of a Social Norms Survey. *International Journal of Behavioral Development, 45*(3), 256–268.

Geertz, C. (1973). *The interpretation of cultures.* New York, NY: Basic Books.

Goto, S., Ando, Y., Huang, C., Yee, A., & Lewis, R. (2010). Cultural differences in the visual processing of meaning: Detecting incongruities between background and foreground objects using the N400. *Social Cognitive and Affective Neuroscience, 5,* 242–253.

Hedden, T., Ketay, S., Aron, A., Markus, H., & Gabrieli, J. (2008). Cultural influences on neural substrates of attentional control. *Psychological Science, 19,* 12–17.

Islam, F., Khanlou, N., & Tamim, H. (2014). South Asian populations in Canada: migration and mental health. *BMC Psychiatry, 14,* 154.

Ito, K., Masuda, T. & Li, M. (2013). Agency and facial emotion judgment in context. *Personality and Social Psychology Bulletin, 39,* 763–776.

King, D.J., Hodgekins, J., Chouinard, P.A., Chouinard, V.A., & Sperandio, I. (2017). A review of abnormalities in the perception of visual illusions in schizophrenia. *Psychonomic Bulletin & Review, 24,* 734–751.

Kirmayer, L., Weinfeld, M., Burgos, G., Du Fort, G., Lasry, J., & Young, A. (2007). Use of health care services for psychological distress by immigrants in an urban multicultural milieu. *The Canadian Journal of Psychiatry, 52,* 295–304.

Kitayama, S. & Cohen, D. (2019). *Handbook of cultural psychology (Second edition).* New York: Guilford Press.

Lee, H. (2019). *Influence of cultural contexts on daily stress experiences: Perception of interpersonal vs. non-interpersonal situations among European Canadian and Japanese Undergraduate Students* [Doctoral Dissertation, University of Alberta]. Education and Research Archive.

Lin, Z., & Han, S. (2009). Self-construal priming modulates the scope of visual attention. *Quarterly Journal of Experimental Psychology, 62*(4), 802–813.

Markus, H., & Kitayama, S. (1991). Culture and the self: Implications for cognition, emotion, and motivation. *Psychological Review, 98*, 224–253.

Masuda, T. (2009). Cultural effects on visual perception. In Goldstein, E.B. (Ed.) *Encyclopedia of Perception* (pp. 339–343). Sage Publications: USA.

Masuda, T., Ellsworth, P., Mesquita, B., Leu, J., Tanida, S., & van de Veerdonk, E. (2008). Placing the face in context: Cultural differences in the perception of facial emotion. *Journal of Personality and Social Psychology, 94*, 365–381.

Masuda, T. & Nisbett, R.E. (2001). Attending holistically vs. analytically: Comparing the context Sensitivity of Japanese and Americans. *Journal of Personality and Social Psychology, 81*, 922–934.

Masuda, T., Russell, M., Chen, Y., Hioki, K., & Caplan, J. (2014). N400 incongruity effect in an episodic memory task reveals different strategies for handling irrelevant contextual information for Japanese than European Canadians. *Cognitive Neuroscience, 5*, 17–25.

Masuda, T., Russell, M., Li, L., & Lee, H. (2019). Perception and cognition. In S. Kitayama, & D. Cohen (Eds.), *Handbook of cultural psychology*. New York: Guilford Press.

Masuda, T., Wang, H., Ishii, K., & Ito, K. (2012). Do surrounding figures' emotions affect judgment of the target figure's emotion?: Comparing the eye-movement patterns of European-Canadians, Asian-Canadians, Asian International Students, and Japanese. *Frontiers in Integrative Neuroscience, 6*, 72. doi:10.3389/fnint.2012.00072.

Matsumoto, D., Kwang, H., & Yamada, H. (2010). Cultural differences in the relative contributions of face and context to judgments of emotions. *Journal of Cross-Cultural Psychology, 43*, 198–218.

Miyamoto, Y., Yoshikawa, S., & Kitayama, S. (2011). Feature and configuration in face processing: Japanese are more configural than Americans. *Cognitive Science, 35*, 563–574.

Mosimann, U.P., Mather, G., Wesnes, K.A., O'Brien, J.T., Burn, D.J., & McKeith, I.G. (2004). Visual perception in Parkinson disease dementia and dementia with Lewy bodies. *Neurology, 63*, 2091–2096.

Na, J., Grossmann, I., Varnum, M., Kitayama, S., Gonzalez, R., & Nisbett, R.(2010). Cultural differences are not always reducible to individual differences. *Proceedings of the National Academy of Sciences, 107*, 6192–6197.

Na, J., & Kitayama, S. (2011). Spontaneous trait inference is culture-specific behavioral and neural evidence. *Psychological Science, 22*, 1025–1032.

Nakamura, Y., Takeuchi, T., Hashimoto, K., & Hashizume, M. (2017). Clinical features of outpatients with somatization symptoms treated at a Japanese psychosomatic medicine clinic. *BioPsychoSocial Medicine, 11*(*1*), 16.

Nisbett, R. (2003). *The geography of thought*. New York, NY: Free Press.

Nisbett, R. & Masuda, T. (2003). Culture and point of view. *Proceedings of the National Academy of Sciences of the United States of America, 100*, 11163–11175.

Razani, J., Burciaga, J., Madore, M., & Wong, J. (2007). Effects of acculturation on tests of attention and information processing in an ethnically diverse group. *Archives of Clinical Neuropsychology, 22*(3), 333–341.

Russell, M., Li, L., Lee, H., Singhal, A., & Masuda, T. (2020). Neural cultural fit: Non-social and social flanker task N2s and well-being in Canada. *Culture and Brain, 8*, 186–206. doi: 10.1007/s40167-019-00089-8.

Russell, M., Masuda, T., Hioki, K., & Singhal, A. (2015). Culture and social judgments: The importance of culture in Japanese and European Canadians' N400 and LPC processing of face lineup emotion judgments. *Culture and Brain, 3*, 131–147.

Russell, M., Masuda, T., Hioki, K., & Singhal, A. (2019). Culture and neuroscience: How Japanese and European Canadians process social context in close and acquaintance relationships. *Social Neuroscience, 14*(4), 484–498. doi: 10.1080/17470919.2018.1511471.

Ryder, A., Bean, G., & Dion, K. (2000). Caregiver responses to symptoms of first-onset psychosis: A comparative study of Chinese-and Euro-Canadian families. *Transcultural Psychiatry, 37*(2), 255–266.

Ryder, A., Yang, J., Zhu, X., Yao, S., Yi, J., Heine, S., & Bagby, R. (2008). The cultural shaping of depressive symptoms: Somatization and psychologization in China and North America. *Journal of Abnormal Psychology, 117*, 300–313.

Segall, M.H., Campbell, D.T., & Herskovits, M.J. (1966). *The influence of culture on visual perception* (pp. 174–184). Indianapolis: Bobbs-Merrill.

Senzaki, S., Masuda, T., & Ishii, K. (2014). When is perception top-down and when is it not? Culture, narrative, and attention. *Cognitive Science, 38*(7), 1493–1506.

Tsai, J., Simeonova, D., & Watanabe, J. (2004). Somatic and social: Chinese Americans talk about emotion. *Personality and Social Psychology Bulletin, 30*(9), 1226–1238.

Varnum, M., Grossmann, I., Kitayama, S., & Nisbett, R.E. (2010). The origin of cultural differences in cognition: The social orientation hypothesis. *Current Directions in Psychological Science, 19*, 9–13.

Varnum, M., Na, J., Murata, A., & Kitayama, S. (2012). Social class differences in N400 indicate differences in spontaneous trait inference. *Journal of Experimental Psychology: General, 141*, 518–526.

White, G. (1982). The role of cultural explanations in 'somatization' and 'psychologization'. *Social Science & Medicine, 16*(16), 1519–1530.

Yoo, S. & Skovholt, T. (2001). Cross-cultural examination of depression expression and help-seeking behavior: A comparative study of American and Korean college students. *Journal of College Counseling, 4*(1), 10–19.

Zhou, X., Min, S., Sun, J., Kim, S., Ahn, J., Peng, Y., … & Ryder, A. (2015). Extending a structural model of somatization to South Koreans: Cultural values, somatization tendency, and the presentation of depressive symptoms. *Journal of Affective Disorders, 176*, 151–154.

6

CULTURE AND PERCEPTION

Sumita Chatterjee

Positionality statement

At my current point in life, I am afforded many privileges – e.g., having a high level of education from American and British institutions, having an American nationality, fluency in English, mobile freedom, etc. – while simultaneously, I feel vulnerable to social inequalities related to the constructed meaning ascribed to some of my perceived characteristics – e.g., being perceived as a woman of color. Over the course of my life, my positionality has shifted with changes in my physical and cultural environments, social structure, and intellectual exposure. This has allowed me to gain broader view-points; however, I cannot deny that my privileges and knowledge biases may render me blind to important nuances in my analysis. I therefore welcome constructive criticism in this topic and beyond; I hope to nourish open and empathetic dialog to work through destabilizing emotional responses that are often difficult to articulate.

The following describes milestones that have been recorded in the past; however, I say with much prudence, that the works of many of these individuals, for a variety of intersecting reasons, have gained the loudest voice, and continue to be chronicled as the so-called "seminal works" in this field. That is to say that I encourage the reader to take in the following information with a critical (as opposed to a neutral) eye so as to not inadvertently blindside (pun intended) oneself into a type of "socio-political visual agnosia"; to pursue their own investigations if only to help efforts to shine a light on the diverse forgotten many that have also contributed greatly to the discoveries in the clinical expressions of visual perception.

Introduction

The impact of contrasting societal structures on visual perception is well re-cognized within the academic community. A substantial body of research has

DOI: 10.4324/9781003051497-8

attempted to describe, characterize, and understand apparent cultural differences in perceptual processes, but have primarily framed such differences as the analytical style of the individualistic West versus the holistic style of the collectivist East. This essentialist take on people carries a heavy history of "othering" which, if crystalized through biological discourse, not only perpetuates discriminatory ideas of what certain groups of individuals are able to do, but also sets academic research along a groove of thought that may occlude the researcher to a diversity of innovative questions, or render certain ideas as "irrelevant" or "uninteresting". However, it is undeniable that there is something about culture that guides our visual perception. Before turning to this literature, I will first explore a model of vision which proposes a mechanism of basic visual processing that could explain how cultural experience might impact visual perception.

A model for vision

A recent framework for understanding vision proposed by Li Zhaoping (2019) provides a useful account of visual systems, including a mechanism by which aspects of cultural experience may influence perception. The model emphasizes the selection of visual input in the process of visual processing, focusing on evidence that this process starts in the primary visual cortex (V1). Zhaoping suggests that encoded information coming from the retina and lateral geniculate nucleus creates a "saliency map" in V1 from which small proportions of information are sent forward to other brain areas. For example, it is argued that the human retina receives about 100 MB/ second of visual information but, as a result of efficient encoding, sends about 1 MB/second of visual data to V1. Visual attention then selects a tiny fraction (the equivalent of 40 bits/second) for further conscious processing (Zhaoping, 2019).

From here, two main feedback loops ensue. One is a selection process involving eye movements guided by top-down and bottom-up processing. Top-down processing shifts one's gaze to align task- or goal-relevant objects with the center of the eye, or fovea, so as to "see" with greater detail, whereas bottom-up processing is informed by peripheral vision influencing where to "look" next (Zhaoping, 2019). In other words, the visual system directs eye movements by integrating low-resolution peripheral information and high-resolution foveal information, with our goals, past experiences, and knowledge of the environment (Castelhano et al., 2009; Nuthmann, 2013). This, in turn, continues to update the V1 saliency map.

The second feedback loop involves other cortical areas using internal knowledge of the visual world to synthesize representations of decoded visual inputs coming from V1. The synthesized representations are then relayed back to V1 to match the representations with the "raw" visual inputs. Each representation is then reinforced or weakened depending on the quality of the match, and alternative representations continue to compete with each other for our ultimate perception (for a more in-depth review, see Zhaoping, 2019). These two feedback loops continue to inform each other; simultaneously, our "goals, past experiences, and knowledge of the environment", and "internal knowledge of the visual world" are imbued by the cultural ideas propagated, in part, by our society. However, the precise aspects of culture that influence these, and the mechanism by which this happens is not established. Furthermore, "culture" is a term that comes with much controversy.

Culture and visual perception

Many have tried to define "culture" in an attempt to break away from earlier oppressive ideas of "civilized" versus "barbaric" cultures (Arnold, 1869); however, in the world of perception, cultural differences in perceptual abilities have been attributed to thinking styles encouraged by various characteristics of a social structure. Early studies indicated that aspects of perceptual experience are influenced by exposure to particular environments, something that can be demonstrated by different responses to perceptual illusions such as the Müller-Lyer illusion or the Sander parallelogram (Segall et al., 1999). One popular characterization of the influence of culture on cognition is the analytical style of the individualistic West, and the holistic style of the collectivist East. Masuda and Nisbett's (2001) classic study compared the performance of Japanese and American participants on an underwater scene description task. It was found that Americans emphasized focal objects (the large, brightly colored, rapidly moving objects) while the Japanese reported more information about the background (e.g., rocks, color of water, small non-moving objects) compared to the Americans. Participants also did a memory task involving viewing pictures of a single animal against a background scene. Later, they made old/new recognition judgments for animals in scenes – sometimes the focal animal was shown against the original background; other times the focal animal was shown against a new background. Japanese and Americans were equally accurate in recognizing the focal animal in its original background. However, Americans were more accurate when the animal was displayed against a new background. Masuda and Nisbett concluded that the Japanese encoded the scenes more holistically than the Americans, binding information about the objects with the backgrounds, so that an unfamiliar new background adversely affected retrieval of the familiar animal.

With developments in eye tracking and brain imaging technologies (e.g., EEG, fMRI), there has been a newfound interest in using them to understand these behavioral differences at a kinematic and neurological level (Chua et al., 2005; Paige et al. 2017; Wang et al., 2014). Chatterjee (2021) conducted a systematic review of studies up until 2019 that had utilized a range of technologies to investigate culture and object/scene perception. A total of 38 articles were included, 22 of which used eye-tracking. Ten eye-tracking studies involved viewing scenes and some found East-West differences in eye movements. For example, Chua et al. (2005) found that Americans fixated more on focal objects than did Chinese participants, and the Americans tended to look at the focal object more quickly. Additionally, the Chinese made more saccades to the background than did the Americans. However, this finding was not replicated in other studies. There were variations in study design that may have accounted for differences in results – it appeared that tasks involving "top-down" processing (i.e., the participant has to actively engage attention to the scene) were more likely to produce cultural differences than passive viewing conditions. In addition, the saliency of the focal object varied between studies which played a factor in detecting cultural difference – the more salient the focal object, the more likely a cultural difference was detected. Twelve studies used other paradigms (Change Blindness, Saccade, Narrative Construction with Motion Video, Reading Direction) with variable findings, though again the East-West differences were only seen when tasks required more top-down attentional processing.

Six studies used EEG to investigate cultural differences in scene perception. Nine different EEG waves were measured between the six papers, eight of which showed the presence of a cultural difference pertaining to object processing. However, only the P3 Target (indicating neural activity involved in detecting rare, but meaningful events) and P3 Novelty (the amount of attention given to a stimulus that appears on occasion but is not what is actively sought out for) EEG waves were investigated in more than one study, and the results were inconsistent despite both studies using the three-stimulus oddball paradigm. The results of the EEG studies were partially congruent with ten studies that investigated cultural differences in scene/object perception using fMRI. The fMRI studies consistently showed a cultural difference in activation in areas known to be involved in object processing as contained units, and not part of a larger context or scene; however, whether background processing differed between cultures remained unclear.

It seems then that cultural differences in scene perception are apparent under some circumstances but the precise conditions under which these differences emerge need further clarification. All studies included in the systematic review, with the exception of one, utilized complex images – i.e., scenes or patterns of geometric shapes – and thus, it is not clear how these results would extend to cultural differences in visual perception of single objects, which are a majority of the stimuli used in neuropsychological tests of visual perception. What appears to be clear is that the differences primarily operate at the level of attention rather than underlying basic perceptual processing, but as Zhaoping's model suggests, attention and selection processes are operating very early in the visual processing system, in V1. In other words, cultural factors may influence what participants attend to in a scene, which then impacts how information is processed and subsequently recalled in memory tasks. The key point is that differences in visual perception could influence task performance, depending on the nature of the task.

It should also be noted that these studies mostly used American and Chinese or Japanese participants as the exemplars of the "West" and "East". A few studies have shown that this characterization is not confined to just North America and East Asia (Grossmann & Kross, 2010; Kitayama et al., 2006; Knight & Nisbett, 2007; Varnum et al., 2008). However, more importantly, the dichotomization of the globe into West versus East, individualist versus collectivist, and so on ignores the socio- and geopolitical history of othering/oppression that the context of such vocabulary carries (Hall, 1992; Said, 1978). It also ignores the spectrum that exists between collectivism, individualism, and beyond, thus hindering space for uniqueness. For example, many studies have shown that describing a country such as India as being part of the "East" (Chakkarath, 2010) and perceiving it as a collectivist culture (Country Comparison – Hofstede Insights, 2018) is a mischaracterization. These studies argue that India, being very diverse within itself, is a spread between collectivism and individualism with certain areas tending more toward collectivism, others more toward individualism, and some places showing both individualistic and collectivistic characteristics existing symbiotically (Jha & Singh, 2011; Sinha et al., 2001). Furthermore, many individuals, along with many other cultures, do not fit the historical mold of "West" or "East"; using this dualistic lens limits the vocabulary for questioning/researching culture and visual perception, and can result in pigeon-holing those outside the bounds of "East" and "West" into the framework (Duan et al., 2016).

A more prudent approach to culture and visual perceptual research would be to take a more anthropological approach to culture, and to incorporate models and methods implemented within a culture/social studies, instead of disguising essentialist assumptions like individualism versus collectivism, as facts. For example, some models of culture see cultural ideas – e.g., religious beliefs, physical environment, genetics/biology, and social structure to be four separate "categories" that are highly interlinked (Brown, 2008). In addition, critical feminism expands upon social structures to consist of intersecting characteristics that carry power dynamics relative to each other – e.g., race, ethnicity, gender, religion, physical ability, power, caste, language, education quality, class, etc. These can influence an individual's position in society and therefore might shape how others in the same society see them/how they see and maneuver themselves within that space. It acknowledges a wider spread of characteristics well beyond collectivism and individualism, while also acknowledging that an individual's position in society can change over the course of time; it allows space for the unique experiences of individuals, and steps away from essentialist dichotomies which greatly overgeneralize large groups of people and perpetuate stereotypes (Qin, 2004). Using these models in research involves incorporating qualitative methods – e.g., ethnography – along with exploring the use of more flexible quantitative methods – e.g., Bayesian modeling – in order to gain a fuller, more nuanced, and more representative understanding of the specific aspects of culture that may be influencing visual perception. This then has major implications for the process of neuropsychological assessment, which aims to identify impairments in visual perceptual processes arising in the context of neurological injury and disease. Furthermore, considering indigenous ecological knowledge/conceptualizations of various clinical expressions of visual perceptual dysfunctions can help improve a clinician's approach to a patient. For example, working with local healers in First Nation communities in Canada, or Ayurvedic healers in India might help neuropsychological clinicians understand how to frame and co-create treatments that respect the lens through which patients from those communities view their difficulties while helping to build a greater network of trust for the patient. Simultaneously, understanding how different knowledge systems conceptualize the same clinical expression of visual dysfunction increases our vocabulary, thus opening us up to new awarenesses, and broadening our scope for alternative lines of thought and questioning. This widens the horizon for more inclusive research, flexible thought, and a more all-encompassing clinical practice.

It is important to recognize that knowledge systems are not neutral; they carry a power dynamic (Knopf, 2015). Taking a moment to step back and think about the origins of our knowledge system can give us perspective on how we may or may not be contributing to keeping such power structures alive. For example, if we find that our knowledge base resides in a more colonial school of thought, it is useful to take steps toward decolonizing our minds by collaborating with experts of different knowledge systems with open curiosity and humility, as opposed to approaching such experts with resistance to the possibility of changing/expanding our own current knowledge base. For example, feeling like we do not need to change ourselves because the experts of the other knowledge systems will do the necessary job, or holding onto our way of thinking because of a conscious/subconscious notion that our way of thinking is best suited; both modes shift responsibility onto the shoulders of the oppressed to convince those in the power of their legitimacy (Knopf, 2015). We must also hold ourselves accountable to not solely rely on these experts to teach us, but to also take responsibility in teaching ourselves.

Much of the globally accepted/standard discourse and treatment for visual perception and its disorders are European in origin. It is important, however, to acknowledge that other scientific traditions exist that may have fallen by the wayside during colonial periods, and which are being reconsidered as scholars and activists challenge Aristotelian "Western" science and methods (Haraway, 1988; Kraus, 2009; Smith, 1999). Many branches of science, including neuropsychology, are beginning to "decolonize" (e.g., Cagigas et al., 2021), and it will be important to carefully consider the extent to which methods and interpretation of data from studies of perception, including studies of cultural influences on perception, reflect the influence of a colonial history.

Neuropsychological assessment of visual perception

A variety of standardized visual perceptual neuropsychological batteries were created during the second half of the 20th century. These tests are designed to target the different facets of perception, however, they were typically made to mimic stimuli from the environment in which they were developed. Therefore, the "expected" normative performance is based on the responses from people originating from that specific place. Individuals from different environments may perform below the "normal range" suggesting cognitive impairments, but a clinician may not know how much to attribute this apparent impairment to the individual or to the inherent bias of the test. Thus, these tests are rendered less effectual in detecting cognitive dysfunction (Agranovich & Puente, 2007).

Many of the commonly used neuropsychological assessments, including tests of perception, have been developed in the United States or the United Kingdom and are, therefore, designed for English-speaking, Western, populations. For example, the Visual Object & Space Perception battery (VOSP; Warrington & James, 1991) is a battery that consists of eight subtests: four for object perception and four for space perception. When this battery was created, residents of England were used in the collection of normative data. Thus, stimuli were created around what was expected of British people. When presenting this battery to different populations, results vary across cultural groups (Herrera-Guzmán et al., 2004). In India, for example, performance on the object perception subtests was considerably lower than the expected normative score (Chatterjee, 2021; Dutt et al., 2016). Chatterjee (2021) specifically investigated self-construal, familiarity, and eye movements as driving "cultural" factors for performance disparities between the British and Indians on the VOSP Silhouettes subtest (identifying objects from foreshortened silhouettes of common animals and objects). A significant difference in overall performance was driven by differences for certain objects. On tests of object perception (and object naming), the most obvious explanation for performance differences between cultures is simply a lack of familiarity with the test stimuli. However, the difference in performance on the subtest between the Indians and the British could not be explained by object familiarity, eye movements, or self-construal, implying that the driving factor(s) may not have anything to do with broad differences in perception but rather differences specific to particular items. Chatterjee (2021) had speculated that for certain items, familiarity may have played a role – e.g., a corkscrew is not a culturally relevant object in India; therefore, the low accuracy rate on that object would be expected. For certain other items, it may not have been due to the participants' familiarity with the object per se, but rather an unfamiliarity with the representation of that object – e.g., the rabbit appears to be a silhouette of the Easter chocolate rabbits typically sold in the UK, but nothing similar is present in India. Similarly, Indians performed poorly on all of the four-legged animals compared to the norm: Bear, Rhinoceros, Sheep, and Cow. This might have to do with the great diversity of four-legged animals that exist in India and therefore the silhouetted objects may not be distinct enough for Indians to identify them correctly. It should be noted that Chatterjee's British participants also performed poorly on the four-legged animals compared to the norm – the Indians and the British performed comparably for the Sheep and the Cow – which might suggest that the general representation of these items may be outdated. But for certain items, perhaps a perceptual difference could explain the disparity in accuracy, since these items are very pervasive in the Indian context – e.g., the Frog, Snail, and Sunglasses. Nonetheless, no one specific factor could explain the overall performance difference; rather it could be that different factors explain different portions of the items in the subtest. It should be noted that samples from the UK and India carried a positionality as well – i.e., English speaking, high education, high caste, mainly individuals from Scotland, Kolkata, and Bangalore, etc. This is not to say that the data are not informative but that many other social factors should be considered when making representative conclusions of a population. For example, typically, schools taught in English are very expensive and are mainly accessible to those belonging to a higher socio-economic class. Thus the resources available for global exposure – e.g., ability to travel – and educational support – e.g., access to private tutors, have parents who are also highly educated, etc. – may be greater. One can then argue that if people belonging to this section of society are performing poorly, then it can be assumed that those with less access to resources and global exposure would perform even worse. Though this may be a safe assumption, to leave it as an assumption, or worse, begin claiming it as a truth without experimental verification would be a grave oversight.

In response to the fact that performance differences between cultures are observed on many neuropsychological tests, including tests of perception, many countries have started to adapt these assessments to better suit their populations (Albonico et al., 2017; Fernández & Fulbright, 2015). This movement in increasing awareness toward creating more culturally compatible tests came with greater vigor starting from the end of the 20 century (Puente & Agranovich, 2003; Rao et al., 2004) and with it came the necessity to take a broader look at what factors are influencing performance on cognitive tests. Though these batteries represent steps toward the improvement of neuropsychological testing, they still do not cover all aspects of cognition including visual perception. In fact, aside from the VOSP, little to no studies have investigated the cultural compatibility of other full visual perceptual neuropsychological batteries. As a result, unadapted assessments are still being used to fill in the gaps.

Like many other domains, tests of visuospatial skills can be influenced by educational experience. For example, one of the most commonly used visuospatial screening tests is the Clock Drawing Test, which is included in cognitive test batteries designed to be used in multicultural contexts such as the Cross-Cultural Neuropsychological Test Battery (Nielsen et al., 2018). But the Clock Drawing Test is significantly affected by education (Crombie, 2021) and so caution is needed in interpreting performance in people with low levels of education.

Implications for clinical practice

As with all other domains of cognition, cultural experiences influencing performance on perceptual tests highlight the importance of careful consideration of the assessment process itself. Particularly for clinicians in contexts where locally-developed tests do not exist, or for clinicians testing a person from a different cultural background than themselves. As also noted in other chapters in this book, the solution is often not simply a case of collecting new "local" norms for a test imported from elsewhere – it may be sufficient, but new studies of validity will be required before one has confidence that the test is working for its intended purpose. These findings suggest that "development from scratch" should also be considered i.e., developing a test using data on how perception functions and is impaired by neurological conditions and injury, but framed within the new context.

References

Agranovich, A.V., & Puente, A.E. (2007). Do Russian and American normal adults perform similarly on neuropsychological tests? Preliminary findings on the relationship between culture and test performance. *Archives of Clinical Neuropsychology, 22*(3), 273–282.

Albonico, A., Malaspina, M., & Daini, R., (2017). Italian normative data and validation of two neuropsychological tests of face recognition: Benton Facial Recognition Test and Cambridge Face Memory Test. *Neurol Sci, 38*(9),1637–1643.

Arnold, M. (1869). *Culture and anarchy*. New York, Cambridge: Cambridge University Press, 1932.

Brown, M. J. (Ed.). (2008). *Explaining culture scientifically*, Seattle, WA: University of Washington Press.

Cagigas, X., Suarez, P., Díaz-Santos, M., Ikanga, J., Kamalyan, L., & Yáñez, J.J. (2021). Decolonizing neuropsychology. In *Symposium held at the Annual Meeting of the International Neuropsychological Society*, San Diego, 2nd–5th February, 2021.

Castelhano, M.S., Mack, M.L., & Henderson, J.M. (2009). Viewing task influences eye movement control during active scene perception. *Journal of Vision, 9*(3):6, 1–15.

Chakkarath, P. (2010) Internationalizing education and the social sciences: Reflections on the Indian context. In M. Kuhn & D. Weidemann (Eds.), *Internationalization of the social sciences: Asia – Latin America – Middle East – Africa – Eurasia*, Bielefeld, Germany: transcript (pp. 87–114).

Chatterjee, S. (2021). *Differences in object perception: A comparison of Indian and British participants on scene and silhouetted object perception tasks* (Unpublished doctoral dissertation), University of Glasgow, Glasgow.

Chua, H.F., Boland, J.E., & Nisbett, R.E. (2005). Cultural variation in eye movements during scene perception. *Proceedings of the National Academy of Sciences of the United States of America, 102*(35), 12629–12633.

Crombie, M. (2021). *Examination of the impact of education on cognitive screening tests*. D Clin Psy thesis, University of Glasgow.

Duan, Z.H., Wang, F.X., & Hong, J.Z. (2016). Culture shapes how we look: Comparison between Chinese and African university students. *Journal of Eye Movement Research, 9*(6), 10.

Dutt, A., Rao, P.S., Kapur, N., Bhargava, P., Rapport, L.R., Millis, S., Peña Casanova, J., Kosmidis, M.H., & Ghosh, A. (2016). Complex visual processing differences across cultures: Evidence from a cross-cultural study of the Visual Object and Space Perception Battery. *Paper presented at the Mid-Year Meeting of the International Neuropsychological Society*, London, 6th–8th July, 2016.

Fernández A.L., & Fulbright R.L., (2015). Construct and concurrent validity of the Spanish adaptation of the boston naming test. *Applied Neuropsychology: Adult, 22*,355–362.

Grossmann, I., & Kross, E. (2010). The impact of culture on adaptive versus maladaptive self-reflection. *Psychological Science, 21*(8), 1150–1157.

Hall, S. (1992). The West and the rest: Discourse and power. In *Formations of modernity*. Open University.

Haraway, D. (1988). Situated knowledges: The science question in feminism and the privilege of partial perspective linked references are available on JSTOR for this article. *Feminist Studies, 14*(3), 575–599.

Herrera-Guzmán, I., Peña-Casanova, J., Lara, J.P., Gudayol-Ferré, E., & Böhm, P., (2004). Influence of age, sex, and education on the Visual Object and Space Perception Battery (VOSP) in healthy normal elderly population. *Clinical Neuropsychology, 18*, 385–394.

Hofstede Insights. 2018. *Country comparison – Hofstede insights* [online]. Available at: https://www.hofstede-insights.com/country-comparison/thailand/

Jha, S.D., & Singh, K. (2011). An analysis of individualism-collectivism across Northern India. *Journal of the Indian Academy of Applied Psychology, 37*(1), 149–156.

Kitayama, S., Ishii, K., Imada, T., Takemura, K., & Ramaswamy, J. (2006). Voluntary settlement and the spirit of independence: Evidence from Japan's "northern frontier". *Journal of Personality and Social Psychology, 91*(3), 369–384.

Knight, N., & Nisbett, R.E., (2007) Culture, class, and cognition: Evidence from Italy. *Journal of Cognition and Culture. 7*, 283–291.

Knopf, K. (2015). The turn toward the indigenous: knowledge systems and practices in the academy. *Amerikastudien / American Studies, 60*(2/3), 179–200.

Kraus, C. (2009). What is the feminist critique of neuroscience? *Neuroethics, 5*, 247–259.

Masuda, T., & Nisbett, R.E. (2001). Attending holistically versus analytically: Comparing the context sensitivity of Japanese and Americans. *Journal of Personality and Social Psychology, 81*(5), 922–934.

Nielsen, T. R., Segers, K., Vanderaspoilden, V., Bekkhus-Wetterberg, P., Minthon, L., Pissiota, A., Bjørkløf, G. H., Beinhoff, U., Tsolaki, M., Gkioka, M., & Waldemar, G. (2018). Performance of middle-aged and elderly European minority and majority populations on a Cross-Cultural Neuropsychological Test Battery (CNTB). *The Clinical Neuropsychologist, 32*(8), 1411–1430. 10.1080/13854046.2018.1430256.

Nuthmann, A. (2013). How do the regions of the visual field contribute to object search in real-world scenes? Evidence From eye movements. *Journal of Experimental Psychology. Human Perception and Performance, 40*, 342–360.

Paige, L.E., Ksander, J.C., Johndro, H.A., & Gutchess, A.H. (2017). Cross-cultural differences in the neural correlates of specific and general recognition. *Cortex, 91*, 250–261.

Puente, A., & Agranovich, A. (2003). *The cultural in cross-cultural neuropsychology* (pp. 321–332).

Qin, D. (2004). *Toward a Critical Feminist Perspective of Culture and Self, 14*(2), 297–312.

Rao S.L.,. Subbukrishna D.K., & Gopukuar K., (2004). Nimhans neuropsychology battery – 2004. *NIMHANS Publication No. 60*. Bangalore: NIMHANS.

Said, E.W. (1978). *Orientalism*. New York: Pantheon Books.

Segall, M.H., Dasen, P.R., Berry, J.W., & Poortinga, Y.H. (1999). *Human behavior in global perspective: An introduction to cross-cultural psychology* (Second revised edition). Boston: Allyn & Bacon.

Sinha, J.B.P., Sinha, T.N., Verma, J., & Sinha, R.B.N. (2001). Collectivism coexisting with individualism: An Indian scenario. *Asian Journal of Social Psychology, 4*: 133–145.

Smith, L.T. (1999). *Decolonizing methodologies: Research and indigenous peoples*. Zed Books Ltd & University of Otago Press.

Varnum, M.E.W., Grossmann, I., Katunar, D., Nisbett, R.E., & Kitayama, S. (2008). Holism in a European cultural context: Differences in cognitive style between Central and East Europeans and Westerners. *Journal of Cognition and Culture, 8*, 321–333.

Wang, K., Umla-Runge, K., Hofmann, J., Ferdinand, N.K., & Chan, R.C.K. (2014). Cultural differences in sensitivity to the relationship between objects and contexts: evidence from P3. *Neuroreport, 25*(9), 656–660.

Warrington, E.K., & James, M. (1991). *The visual object and space battery perception*. Bury St Edmunds: Thames Valley Company.

Zhaoping, L. (2019). A new framework for understanding vision from the perspective of the primary visual cortex. *Current Opinion in Neurobiology, 58*, 1–10.

7

THE INFLUENCE OF CULTURE ON MEMORY

Bernice A. Marcopulos and Kara Eversole

Introduction

Memory is the most intensely studied cognitive function in cognitive neuroscience and clinical neuropsychology. It is one of the most common presenting problems in referrals for neuropsychological assessment (Rabin et al., 2016). Do the theoretical assumptions currently governing our comprehension of normal and abnormal memory processes in the brain hold true across cultures? Are neuropsychological tests which measure memory valid across cultures? In this chapter, we will briefly review the underlying theoretical bases for modern clinical memory assessment. We will discuss the ontogeny of memory development and summarize studies on culture and memory in non-clinical populations. Finally, we will make recommendations for furthering our understanding of memory functioning across cultures and for developing the best methods for assessing and diagnosing memory difficulties in a culturally relevant manner.

A brief history of current memory constructs and testing methods

Philosophers and scientists have studied human memory for centuries. The scientific study of memory as a function of the human mind dates as far back as the early 1800s and has multidisciplinary roots, beginning in philosophy before expanding into psychology and biology (for reviews, see Bower, 2000; Squire, 2004). Memory testing paradigms such as list learning, paired associates, and memory for figures which are commonly used in clinical neuropsychology as well as cognitive neuroscience were introduced over 100 years ago. In Germany, Ebbinghaus (1885) devised experimental methods using syllable list learning (serial learning) to study his own memory. In 1894 in the United States, Mary Calkins

DOI: 10.4324/9781003051497-9

developed the paired associates paradigm (Madigan & O'Hara, 1992). In the early 1900s, the Swiss psychologist Edouard Claparède created a word list to study memory in his patients which was adapted by his doctoral student Andre Rey to become the Rey Auditory Learning Test (Boake, 2000). In England, Bartlett (1932) described using faces, stories, and signs in his experiments on memory in his classic monograph *Remembering*.

The information processing approach to the study of attention and memory was introduced in the mid-20th century (Broadbent, 1958). In 1968, Atkinson and Shiffrin proposed one of the first comprehensive memory models in psychology in their classic paper entitled, "Human Memory: A Proposed System and Its Control Processes". They theorized "the modal model" which posits memory as a combination of three structural components: the sensory register, the short-term store (i.e., short-term memory or working memory), and the long-term store (i.e., long-term memory). They also proposed that memory involves processes of storage, search, and retrieval within the multi-tiered organization. This model has endured over the past 50 years and has greatly influenced the research in memory and cognition that followed its publication (Malmberg et al., 2019). Many of the important memory models that followed (e.g., Baddeley & Hitch, 1974; Craik & Lockhart, 1972; Tulving, 1972), were influenced by the work of Atkinson and Shiffrin.

Memory tests developed for clinical use usually measure the degree to which an individual is able to retain and retrieve information, based on the cognitive models and theories described above. These theoretical models and methods for studying memory were developed primarily in the so-called WEIRD (Western, Educated, Industrialized, Rich, and Democratic, Henrich et al., 2010) countries. This Western perspective has directly influenced how cognitive neuroscientists and clinical neuropsychologists understand and conceptualize memory. When culture is studied as a variable affecting test performance it shows members of the test developer's cultural group (typically WEIRD) usually achieve better test results (Ardila, 2020). As we will see below, the assumptions inherent in "Western" theories of memory organization, structures and processes may not always transport easily across cultures.

Studies of memory and culture

Psychologists have been interested in the cultural influences on cognitive functions since at least the mid-20th century (Triandis, 2007). Luria and Vygotsky (1976, 1978) conducted studies on the cultural influences on cognition in the Soviet Union in the 1930s. Bartlett's (1932) seminal work recognized that culture affects memory. He remarked that memory was a reconstructive process, strongly influenced by social context. He reported some cross-cultural observations, viewing differences from an ethnocentric lens ("primitive" individuals from Swaziland tend to rely on more "concrete, rote memory" processes, p. 265).

In the 1970s, researchers like Rogoff and Cole published groundbreaking studies deconstructing the factors underlying similarities and differences in cognitive task performance across cultures. A basic assumption was that the fundamental processes of memory are similar across cultures (e.g., encoding, retrieval), but *what* is recalled may differ (Cole et al., 1971). These early researchers recognized that cultural differences in cognitive functions such as memory are based on meaningful tasks practiced by that society and emphasized in their educational systems. They also recognized how difficult it is to investigate "pure" memory processes as they have been embedded in prior social and cultural experiences. According to Wagner (1981), a good place to start in understanding the cultural influences of memory is to study how memory emerges in child development – the ontology of memory (for a more recent review, see Chiao, 2018). "Hard-wired" processes emerge early in neurodevelopment, and then the control systems which are most heavily influenced by the environment and education become influential. Thus, it is important to distinguish between the processes of memory versus content of memory when exploring cultural influences.

Rogoff and Mistry (1985) assert that cultural differences are tied to practical and functional factors. They argued that children and adults practice memory skills in their social contexts. Memory is best conceptualized from a functional and educational standpoint.

"Thus, story recall is embedded in the immediate and the broader social context. The performance of a subject in an experiment cannot be considered a window on some pure aspect of memory functioning; it is inherently grounded in the social situation of the current performance and the situations in which the subject is used to remembering things" (Rogoff & Mistry, 1985, p. 134).

Early cultural research showed that cultural differences might be due to memorization strategies learned in school. Cole et al. (1971) compared American schoolchildren with children in Liberia and found that the American children were more likely to use semantic clustering strategies to organize the material to be remembered. Western education teaches children how to organize items based on taxonomy rather than how the object is used. Non-western (e.g., Asian) groups tend to organize material in terms of relational characteristics and function (Chiu, 1972; Ji et al., 2004). This cultural difference in categorical organization strategies is even more pronounced in older adults (Gutchess et al., 2006).

Not only the type of education but also the amount of education affects memory test performance. Individuals across cultures who have fewer years of formal schooling typically perform lower on common memory testing paradigms such as paired associates (Hall, 1972), and list learning (e.g., Cole et al., 1976). Glymour and Manly found that influences of early education and level of literacy impacted memory testing performance later in life (Glymour et al., 2008; Manly et al., 2003). Variance in educational experience across cultures may account for many of the discrepancies in normative data on neuropsychological tests (e.g., Fernández & Marcopulos, 2008; Rosenqvist et al., 2017).

Not all memory tests variables show cultural differences. In their cross-cultural studies of children's cognition, Rogoff (1981) and Wagner (1981) found inconsistent findings on recognition tasks. Presumably, recognition might be more equitable since it does not require the use of mnemonic strategies, but other studies have not supported this hypothesis. For instance, non-Western children perform as well, or even better on memory for spatial arrangement and location (for review, see Rogoff & Mistry, 1985). Spatial memory tasks may not require deliberate encoding strategies compared to verbal tasks. As reviewed by Rogoff and Mistry, Western and non-Western participants alike excel on memory tasks using familiar materials but perform poorly using non-familiar materials to be remembered. Additionally, object memory tests (e.g., Fuld Object Memory Evaluation and Common Objects Memory Test) have repeatedly demonstrated clinical utility and cross-cultural applicability. Both of these tests have been shown to accurately distinguish clinical groups across ethnically diverse populations (Kempler et al., 2010; Loewenstein et al., 1995).

In addition, to cross-cultural differences due to educational systems and socialization, cultural differences in memory test performance may be due to the cultural complexity of the test. The cultural complexity of a cognitive test refers to how much implicit and explicit cultural context from the culture of the test author is present (Van de Vijver & Poortinga, 1992). This can create a bias in the test when comparing individuals from the same culture as the test author to those from another culture. Thus, the cultural complexity or cultural loading of a test can be a better explanation for differences between majority and minority group members (Helms-Lorenz et al., 2003).

Cultural differences in basic cognitive processes

Familiarity with memory test content and emphasis on different encoding strategies learned in school account for many cultural differences as summarized above. However, culture influences memory in more fundamental ways. Cultural influences direct what people attend to in their perceptual field and how they process and interpret this perceptual information (e.g., Nisbett & Miyamoto, 2005; Oyserman & Lee, 2008). The science is uncertain regarding whether the fundamental memory processes and structures differ across culture or whether differences can be reduced to the salience of content (e.g., Leger & Gutchess, 2020). Wang (2021) summarized some of the literature showing how culture alters basic perception, suggesting that basic memory processes are fundamentally different across cultures. Martin and Jones (2012) found that the "observer" perspective was more common in collectivist societies, while a "field" perspective was common in individualist societies. Westerners, who are considered to be from an individualistic society, have a more analytic and field-independent perceptual bias, whereas Asians, who are presumed to be from a more collective society, demonstrate a more holistic, field-dependent perceptual bias. Asians focus more on overall context when viewing and recalling complex visual scenes (e.g., Chua et al., 2005; Nisbett & Masuda, 2003). Western adults are more likely to recall self-focused events whereas Asians recall social interactions (e.g., Wang & Ross, 2005). Wang found that autobiographical memory differs across cultures with Western children recalling specific self-details while children from Asian cultures recall collective details (e.g., Wang, 2004).

Studies of eye-tracking have also shown differences in culture-associated processing styles (i.e., analytic and holistic). In several studies, American participants showed a quicker and longer fixation on foregrounded objects, while Asian participants showed a more balanced fixation that focused on the relationships between objects and backgrounds (Chua et al., 2005; Goh et al., 2009). Differences across cultural groups were also observed in facial perception. East Asian participants tended to focus on the eye region to decipher emotion while Western Caucasian participants showed a more diffuse fixation across facial regions (Jack et al., 2009). When observing faces, Western Caucasians showed a triangular pattern of focus, predominantly fixating on the eyes and partially the mouth, while East Asian participants focused more on the central region of the face (Blais et al., 2008). As with imaging, eye-tracking research has shown mixed results, as Rayner et al. (2007, 2009) found no evidence on cultural effects/differences in eye movement patterns during scene information processing.

The Sapir-Whorf hypothesis posits that an individual's linguistic experience shapes their cognitive processes, including perception and memory (Hunt & Agnoli, 1991; Whorf, 1956). Davidoff (2001) found support for the Whorfian hypothesis in his review of the cross-cultural literature as well as neuropsychological studies of brain-damaged individuals. Studies that are more recent also show the impact of the native language on working memory. Individuals recall information better in the direction of reading for their primary language (left to right or right to left: McCrink & Shaki, 2016). This directionality is also seen in the working memory "white board". Guida and colleagues (Guida et al., 2018) found that Westerners organized their spatial memory "white board" from left to right, whereas Arabic readers organized from right to left. Arabic-speaking illiterates did not show this directionality.

Neuroimaging studies

Neuroimaging studies have also provided evidence that culture influences the neural substrates of cognition, in particular perspective and language processing, attention, and calculation (Han & Northoff, 2008). When listening to music, imaging has shown greater activation in the right ventral medial prefrontal cortex and bilateral precentral gyrus for culturally familiar music, and greater activation in the left cerebellar region, posterior insula, right frontal and angular gyrus, posterior precuneus, and the right middle frontal area extending into the inferior frontal cortex for culturally unfamiliar music (Demorest et al., 2010; Nan et al., 2007). Similarly, differences in imaging were seen during music recall. While the basic processing of auditory stimuli was similar, participants showed greater activation in the left cerebellar region, right angular gyrus, posterior precuneus, and the right middle frontal area extending into the inferior frontal cortex when listening to culturally unfamiliar music, and were more successful in identifying culturally familiar music (Demorest et al., 2010).

Differences in visual processing, particularly when attending to objects, have also been identified and often reflect analytic and holistic styles that tend to be observed across cultural groups (see Han & Northoff, 2008 for a review). When viewing objects, Americans, relative to East Asians, show more activation in areas interrelated to object processing (Gutchess et al., 2006). When viewing faces, East Asians showed more activity in the right fusiform face area (FFA) while Westerners showed bilateral activity in the FFA, and more selectivity in the left FFA (Goh et al., 2010). Changes that are related to age but sensitive to culture have been observed in the lateral occipital cortex (LOC) by Goh et al. (2007). Researchers observed that older Westerners had diminished object processing in the left LOC, while older East Asians saw a reduction in both the right and left LOC (Goh et al., 2007). More recently, Han and Ma (2014) conducted a meta-analysis on 35 MRI studies. This confirmed evidence for cultural differences in neural networks used by Westerners and East Asians in social cognitive processes.

Implications for memory assessment

The research reviewed in this chapter shows how socialization, language, and education shapes how people perceive their world and what they deem important and relevant to remember. As Wang (2021) points out "How we conceive of memory directly influences how we study it" (p. 152). In this chapter, we reviewed the prevailing models of memory function, which reflect Western thought and values. Western culture conceptualizes memory loss as a problem with encoding and retrieval, but other cultures may view memory loss as related to personal habits or life circumstances (Wang, 2021). Memory must be understood within its naturalistic context – what is meaningful and functional within a societal context (e.g., Neisser, 1982). Wang concludes that memory is a "culturally saturated mnemonic system", shaped by numerous dynamic forces, from the macrolevel including global and national influences, family and community and then the micro level, the individual. Memories can be represented in sensory, perception, motor and language systems, all of which are influenced by culture.

Thus, our implicit assumptions that basic memory processes are similar across cultures and that tests can transport across cultures have not withstood closer scrutiny. What "we" think is important to remember and how "we" think is the best way to remember are culturally bound. For instance, one of the most commonly used list-learning tests, the California Verbal Learning Test (Delis et al., 1988) is based on semantic clustering as a preferred recall strategy. However, crosscultural studies show that non-Western cultures do not necessarily organize memory in this way. Some memory tests developed in Western nations have been modified for cross-cultural use and show adequate clinical utility for diagnosing neuropsychological conditions (for a review, see Fernández & Marcopulos, 2019; Fujii, 2017). For example, Arango-Lasprilla and colleagues (2015) adapted and normed the Hopkins Verbal Learning Test – Revised for multiple Spanishspeaking Latin American countries. Lim et al. (2009) adapted a word list test in

French, Mandarin, Malay, and Korean for the cross-cultural assessment using shopping lists adapted to various cultures. Their test produced equivalent results across several cultural groups. Also (as previously mentioned), object memory tests have been validated across cultures, showing specificity and sensitivity across diverse groups. However, developing culturally specific norms for memory tests does not entirely solve the problem of cultural relevance and validity (Fernández & Abe, 2018). The studies showing the impact of education and environment, mediated by culture, on early neurodevelopment suggest that an emic perspective is probably the best solution.

While studies have shown differences in cognitive processes, this body of research can be extended in numerous ways. First, future work should be mindful of the heterogeneity within their samples and limit broad geographical comparisons (e.g., Eastern versus Western individuals) as this can minimize relevant cultural nuances. Second, researchers should also examine cognitive processes in bicultural and bilingual individuals. Prior work has demonstrated that bicultural and bilingual individuals exhibit language-induced frame switches and can be primed by cultural cues in cognitive tasks (Ringberg et al., 2010; Sui et al., 2007). The way information is encoded and recalled may uniquely differ for bicultural and bilingual individuals, as well as individuals at varying levels of acculturation.

References

Arango-Lasprilla, J.C., Rivera, D., Garza, M.T., Saracho, C.P., Rodríguez, W., Rodríguez-Agudelo, Y., & Perrin, P.B. (2015). Hopkins Verbal Learning Test – Revised: Normative data for the Latin American Spanish speaking adult population. *Neurorehabilitation, 37*(4), 699–718. 10.3233/NRE-151286.

Ardila, A. (2005). Cultural values underlying psychometric cognitive testing. *Neuropsychology Review, 15*(4), 185–195. 10.1007/s11065-005-9180-y.

Ardila, A. (2020). Cross-cultural neuropsychology: History and prospects. *RUDN Journal of Psychology and Pedagogics, 17*(1), 64–78. 10.22363/2313-1683-2020-17-1-64-78.

Atkinson, R.C., & Shiffrin, R.M. (1968). Human memory: A proposed system and its control processes. In *Psychology of learning and motivation* (Vol. 2, pp. 89–195). Academic Press.

Baddeley, A.D., & Hitch, G. (1974). Working memory. In *Psychology of Learning and Motivation* (Vol. 8, pp. 47–89). Academic Press.

Bartlett, F.C. (1932). *Remembering: A study in experimental and social psychology.* New York/ London: Cambridge University Press.

Blais, C., Jack, R.E., Scheepers, C., Fiset, D., & Caldara, R. (2008). Culture shapes how we look at faces. *PLoS ONE, 3*(8), 1–8. 10.1371/journal.pone.0003022.

Boake, C. (2000). Edouard Claparède and the auditory verbal learning test. *Journal of Clinical and Experimental Neuropsychology, 22*(2), 286–292.

Bower, G.H. (2000). *A brief history of memory research.* In E. Tulving & F.I.M. Craik (Eds.), *The Oxford Handbook of Memory* (pp. 3–32). Oxford University Press.

Broadbent D.E. (1958). *Perception and Communication.* London: Pergamon.

Chiao, J.Y. (2018). Developmental aspects in cultural neuroscience. *Developmental Review, 50*, 77–89. 10.1016/j.dr.2018.06.005.

Chiu, L. (1972). A cross-cultural comparison of cognitive styles in Chinese and American children. *International Journal of Psychology*, 7, 235–242.

Chua, H.F., Boland, J.E., & Nisbett, R.E. (2005). Cultural variation in eye movements during scene perception. *Proceedings of the National Academy of Sciences*, *102*(35), 12629–12633. 10.1073/pnas.0506162102.

Cole, M. (1988). Cross-cultural research in the sociohistorical tradition. *Human development*, *31*(3), 137–157.

Cole, M., Gay, J., Glick, J., & Sharp, D.W. (1971). *The Cultural Context of Learning and Thinking: An Exploration in Experimental Anthropology*. New York, NY: Basic Books.

Cole, M., & Scribner, S. (1977). Developmental theories applied to cross-cultural cognitive research. *Annals of the New York Academy of Sciences*, *285*(1), 366–373.

Cole, M., Sharp, D.W., & Lave, C. (1976). The cognitive consequences of education: Some empirical evidence and theoretical misgivings. *The Urban Review*, *9*(4), 218–233. 10.1007/BF02175468.

Craik, F.I.M., & Lockhart, R.S. (1972). Levels of processing: A framework for memory research. *Journal of Verbal Learning and Verbal Behavior*, *11*(6), 671–684. 10.1016/S0022-5371(72)80001-X.

Davidoff, J. (2001). Language and perceptual categorisation. *Trends in Cognitive Sciences*, *5*(9), 382–387. 10.1016/S1364-6613(00)01726-5.

Delis, D.C., Freeland, J., Kramer, J.H., & Kaplan, E. (1988). Integrating clinical assessment with cognitive neuroscience: construct validation of the California Verbal Learning Test. *Journal of Consulting and Clinical Psychology*, *56*(1), 123.

Demorest, S.M., Morrison, S.J., Stambaugh, L.A., Beken, M., Richards, T.L., & Johnson, C. (2010). An fMRI investigation of the cultural specificity of music memory. *Social Cognitive and Affective Neuroscience*, *5*(2–3), 282–291. 10.1093/scan/nsp048.

Ebbinghaus, H. (1885). *Über das gedächtnis: untersuchungen zur experimentellen psychologie*. Duncker & Humblot.

Fernández, A.L., & Abe, J. (2018). Bias in cross-cultural neuropsychological testing: Problems and possible solutions. *Culture and Brain*, *6*, 1–35. 10.1007/s40167-017-0050-2.

Fernández, A.L., & Marcopulos, B.A. (2008). A comparison of normative data for the Trail Making Test from several countries: Equivalence of norms and considerations for interpretation. *Scandinavian Journal of Psychology*, *49*, 239–246.

Fernández, A.L., & Marcopulos, B.A. (2019). Cross-cultural tests in neuropsychology: A review of recent studies and a modest proposal. In: Koffler, S., Marcopulos, B.A., Mahone, M., Johnson-Green, D., & Smith, G. (Eds.). *Neuropsychology: Science and Practice* (Vol. III). New York: Oxford University Press.

Glymour, M.M., Kawachi, I., Jencks, C.S., & Berkman, L.F. (2008). Does childhood schooling affect old age memory or mental status? Using state schooling laws as natural experiments. *Journal of Epidemiology & Community Health*, *62*(6), 532–537.

Goh, J.O., Chee, M.W., Tan, J.C., Venkatraman, V., Hebrank, A., Leshikar, E.D., Jenkins, L., Sutton, B.P., Gutchess, A.H., & Park, D.C. (2007). Age and culture modulate object processing and object-scene binding in the ventral visual area. *Cognitive, Affective, & Behavioral Neuroscience*, *7*(1), 44–52. 10.3758/CABN.7.1.44.

Goh, J.O.S., Leshikar, E.D., Sutton, B.P., Tan, J.C., Sim, S.K.Y., Hebrank, A.C., & Park, D.C. (2010). Culture differences in neural processing of faces and houses in the ventral visual cortex. *Social Cognitive and Affective Neuroscience*, *5*(2–3), 227–235. 10.1093/scan/nsq060.

Goh, J.O., Tan, J.C., & Park, D.C. (2009). Culture modulates eye-movements to visual novelty. *PLoS ONE*, *4*(12), 1–9. 10.1371/journal.pone.0008238.

Goody, J. (1998). Memory in oral tradition. In P. Fara & K. Patterson (Eds.), *Memory: The Darwin lectures*. Cambridge University Press.

Guida, A., Megreya, A.M., Lavielle-Guida, M., Noël, Y., Mathy, F., van Dijck, J.P., & Abrahamse, E. (2018). Spatialization in working memory is related to literacy and reading direction: Culture "literarily" directs our thoughts. *Cognition, 175*, 96–100.

Gutchess, A.H., Welsh, R.C., Boduroğlu, A., & Park, D.C. (2006). Cultural differences in neural function associated with object processing. *Cognitive, Affective, & Behavioral Neuroscience, 6*(2), 102–109. 10.3758/CABN.6.2.102.

Gutchess, A.H., Yoon, C., Luo, T., Feinberg, F., Hedden, T., Jing, Q., Nisbett, R.E., & Park, D.C. (2006). Categorical organization in free recall across culture and age. *Gerontology, 52*(5), 314–323. 10.1159/000094613.

Hall, J.W. (1972). Verbal behavior as a function of amount of schooling. *The American Journal of Psychology*, 277–289.

Han, S., & Ma, Y. (2014). Cultural differences in human brain activity: A quantitative meta-analysis. *NeuroImage, 99*, 293–300. doi:10.1016/j.neuroimage.2014.05.062.

Han, S., & Northoff, G. (2008). Culture-sensitive neural substrates of human cognition: A transcultural neuroimaging approach. *Nature Reviews Neuroscience, 9*(8), 646–654. 10.103 8/nrn2456.

He, J., & van de Vijver, F. (2012). Bias and equivalence in cross-cultural research. *Online Readings in Psychology and Culture, 2*(2). 10.9707/2307-0919.1111.

Helms-Lorenz, M., Van de Vijver, F.J.R., & Poortinga, Y.H. (2003). Cross-cultural differences in cognitive performance and Spearman's hypothesis: g or c? *Intelligence, 31*(1), 9–29. 10.1016/S0160-2896(02)00111-3.

Henrich, J., Heine, S.J., & Norenzayan, A. (2010). Most people are not WEIRD. *Nature, 466*, 29. 10.1038/466029a.

Henrich, J., Heine, S., & Norenzayan, A. (2010). The weirdest people in the world? *Behavioral and Brain Sciences, 33*(2–3), 61–83. 10.1017/S0140525X0999152X.

Hunt, E., & Agnoli, F. (1991). The Whorfian hypothesis: A cognitive psychology perspective. *Psychological Review, 98*(3), 377.

Iverson, G.L., Brooks, B.L., Ashton, V.L., Johnson, L.G., & Gualtieri, C.T. (2009). Does familiarity with computers affect computerized neuropsychological test performance? *Journal of Clinical and Experimental Neuropsychology, 31*(5), 594–604. 10.1080/138033 90802372125.

Jack, R.E., Blais, C., Scheepers, C., Schyns, P.G., & Caldara, R. (2009). Cultural confusions show that facial expressions are not universal. *Current Biology, 19*(18), 1543–1548. 10.1016/j.cub.2009.07.051.

Ji, L.J., Zhang, Z., & Nisbett, R.E. (2004). Is it culture or is it language? Examination of language effects in cross-cultural research on categorization. *Journal of Personality and Social Psychology, 87*(1), 57. 10.1037/0022-3514.87.1.57.

Kempler, D., Teng, E.L., Taussig, M., & Dick, M.B. (2010). The common objects memory test (COMT): A simple test with cross-cultural applicability. *Journal of the International Neuropsychological Society, 16*(3), 537–545. 10.1017/S1355617710000160.

Leger, K.R., & Gutchess, A. (2020). Cross-cultural differences in memory specificity: Investigation of candidate mechanisms. *Journal of Applied Research in Memory and Cognition*. 10.1016/j.jarmac.2020.08.016.

Lim, Y.Y., Prang, K.H., Cysique, L., Pietrzak, R.H., Snyder, P.J., & Maruff, P. (2009). A method for cross-cultural adaptation of a verbal memory assessment. *Behavior Research Methods, 41*(4), 1190–1200. 10.3758/BRM.41.4.1190.

Loewenstein, D.A., Duara, R., Argüelles, T., & Argüelles, S. (1995). Use of the fuld object-memory evaluation in the detection of mild dementia among Spanish and English-speaking groups. *The American Journal of Geriatric Psychiatry, 3*(4), 300–307. 10.1 097/00019442-199503040-00004.

Luria, A.R. (1976). *Cognitive development: Its cultural and social foundations.* Boston: Harvard University Press.

Madigan, S., & O'Hara, R. (1992). Short-term memory at the turn of the century: Mary Whiton Calkins's memory research. *American Psychologist, 47*(2), 170.

Malmberg, K.J., Raaijmakers, J.G.W., & Shiffrin, R.M. (2019). 50 years of research sparked by Atkinson and Shiffrin (1968). *Memory & Cognition, 47*, 561–574. 10.3758/s13421-01 9-00896-7.

Manly, J.J., Touradji, P., Tang, M.-X., & Stern, Y. (2003). Literacy and memory decline among ethnically diverse elders. *Journal of Clinical and Experimental Neuropsychology, 25*(5), 680–690. 10.1076/jcen.25.5.680.14579.

Martin, M., & Jones, G. V. (2012). Individualism and the field viewpoint: Cultural influences on memory perspective. *Consciousness and Cognition, 21*(3), 1498–1503. 10.1016/j.concog.2012.04.009.

McCrink, K., & Shaki, S. (2016). Culturally inconsistent spatial structure reduces learning. *Acta Psychologica, 169*, 20–26. 10.1016/j.actpsy.2016.05.007.

Nan, Y., Knösche, T.R., Zysset, S., & Friederici, A.D. (2007). Cross-cultural music phrase processing: An fMRI study. *Human Brain Mapping, 29*, 312–328. 10.1002/hbm.20390.

Neisser, U. (1982). Snapshots or benchmarks? In U. Neisser (Ed.), *Memory observed: Remembering in natural contexts* (pp. 43–48), San Francisco: W. H. Freeman.

Nisbett, R. E., & Masuda, T. (2003). Culture and point of view. *Proceedings of the National Academy of Sciences, 100*(19), 11163–11170. 10.1073/pnas.1934527100.

Nisbett, R.E., & Miyamoto, Y. (2005). The influence of culture: holistic versus analytic perception. *Trends in Cognitive Sciences, 9*(10), 467–473. 10.1016/j.tics.2005.08.004.

Oyserman, D., & Lee, S.W.S. (2008). Does culture influence what and how we think? Effects of priming individualism and collectivism. *Psychological Bulletin, 134*(2), 311–342. 10.1037/0033-2909.134.2.311.

Paige, L.E., Ksander, J.C., Johndro, H.A., & Gutchess, A.H. (2017). Cross-cultural differences in the neural correlates of specific and general recognition. *Cortex, 91*, 250–261. 10.1016/j.cortex.2017.01.018.

Park, D.C., & Huang, C.-M. (2010). Culture wires the brain: A cognitive neuroscience perspective. *Perspectives on Psychological Science, 5*(4), 391–400. 10.1177/1745691610374591.

Pedraza, O., & Mungas, D. (2008). Measurement in cross-cultural neuropsychology. *Neuropsychology Review, 18*(3), 184–193. 10.1007/s11065-008-9067-9.

Rabin, L. A., Paolillo, E., & Barr, W. B. (2016). Stability in test-usage practices of clinical neuropsychologists in the United States and Canada over a 10-year period: A follow-up survey of INS and NAN members. *Archives of Clinical Neuropsychology, 31*(3), 206–230. 10.1093/arclin/acw007.

Rayner, K., Castelhano, M.S., & Yang, J. (2009). Eye movements when looking at unusual/weird Scenes: Are there cultural differences? *Journal of Experimental Psychology: Learning, Memory, and Cognition, 35*(1), 254–259. 10.1037/a0013508.

Rayner, K., Li, X., Williams, C.C., Cave, K.R., & Well, A.D. (2007). Eye movements during information processing tasks: Individual differences and cultural effects. *Vision Research, 47*(21), 2714–2726. 10.1016/j.visres.2007.05.007.

Ringberg, T.V., Luna, D., Reihlen, M., & Peracchio, L.A. (2010). Bicultural-bilinguals: The effect of cultural frame switching on translation equivalence. *International Journal of Cross Cultural Management, 10*(1), 77–92. 10.1177/1470595809359585.

Rogoff, B. (1981). Schooling and the development of cognitive skills. In H.C. Triandis & A. Heron (Eds.), *Handbook of cross-cultural psychology* (Vol. 4, pp. 233–294). Boston: Allyn & Bacon.

Rogoff, B., & Mistry, J. (1985). Memory development in cultural context. In M. Pressley & C.J. Brainerd (Eds.), *Cognitive learning and memory in children* (pp. 117–142). New York: Springer. 10.1007/978-1-4613-9544-7_4.

Rosenqvist, J., Lahti-Nuuttila, P., Urgesi, C., Holdnack, J., Kemp, S.L., & Laasonen, M. (2017). Neurocognitive functions in 3-to 15-year-old children: An international comparison. *Journal of the International Neuropsychological Society 23*, 1–14. 10.1017/S1355 617716001193.

Squire, L.R. (2004). Memory systems of the brain: A brief history and current perspective. *Neurobiology of Learning and Memory, 82*(3), 171–177. 10.1016/j.nlm.2004.06.005.

Sui, J., Zhu, Y., & Chiu, C. (2007). Bicultural mind, self-construal, and self- and mother-reference effects: Consequences of cultural priming on recognition memory. *Journal of Experimental Social Psychology, 43*(5), 818–824. 10.1016/j.jesp.2006.08.005.

Thames, A.D., Hinkin, C.H., Byrd, D.A., Bilder, R.M., Duff, K.J., Mindt, M.R., Arentoft, A., & Streiff, V. (2013). Effects of stereotype threat, perceived discrimination, and examiner race on neuropsychological performance: Simple as black and white? *Journal of the International Neuropsychological Society, 19*(5), 583–593. 10.1017/S1355617713000076.

Triandis, H.C. (2007). Culture and psychology: A history of the study of their relationship. In S. Kitayama & D. Cohen (Eds.), *Handbook of cultural psychology* (pp. 59–76). The Guilford Press.

Tseng, W.-S., Matthews, D.B., & Elwyn, T.S. (2004). *Cultural competence in forensic mental health: A guide for psychiatrists, psychologists, and attorneys*. Brunner-Routledge.

Tulving, E. (1972). Episodic and semantic memory. In E. Tulving & W. Donaldson (Eds.), *Organization of memory* (pp. 381–402). Academic Press.

Van de Vijver, F.J.R., & Poortinga, Y.H. (1992). Testing in culturally heterogeneous populations: when are cultural loadings undesirable? *European Journal of Psychological Assessment*, 17–24.

Vygotsky, L.S. (1978). *Mind in society: The development of higher psychological processes*. Cambridge: Harvard University Press.

Wagner, D.A. (1974). The development of short-term and incidental memory: A cross-cultural study. *Child Development, 45*(2), 389–396. 10.2307/1127960.

Wagner, D.A. (1978). Memories of Morocco: The influence of age, schooling, and environment on memory. *Cognitive Psychology, 10*(1), 1–28. 10.1016/0010-0285(78)90017-8.

Wagner, D.A. (1981). Culture and memory development. *Handbook of Cross-Cultural Psychology, 4*, 187–232.

Wang, Q. (2004). The emergence of cultural self-constructs autobiographical memory and self-description in European North American and Chinese children. *Developmental Psychology, 40*(1), 3. 10.1037/0012-1649.40.1.3.

Wang, Q. (2021). The cultural foundation of human memory. *Annual Review of Psychology, 72*.

Wang, Q., & Ross, M. (2005). What we remember and what we tell: the effects of culture and self-priming on memory representations and narratives. *Memory 13*(6), 594–606. 10.1080/09658210444000223.

Whorf, B.I. (1956). *Language, thought, and reality*. Cambridge, MA: MIT Press.

8

LAYERS OF COMPLEXITY: THE INTERPLAY OF CULTURE AND BI/MULTILINGUALISM ON NEUROCOGNITIVE OUTCOMES ACROSS THE LIFESPAN

Adriana M. Strutt and Beatriz MacDonald

Historical perspective and scientific advancement

Cognition is defined as the mental action or process of acquiring knowledge and understanding through thought, experience, and the senses. This definition highlights the importance of an individual's experience that serves as the foundation for mental processes. Language, on the other hand, has been defined as a powerful mechanism of cultural transmission, which relies on symbols, categories, and labels, including specialized terminology (Gelman & Roberts, 2017). Thus, language is intrinsic to the expression of culture, including our sense of self (Banham, 2014; Boroditsky, 2017 November). When considering the interplay of language, culture, and cognition, Chomsky viewed language as separable from cognition (Berwick et al, 2013) and some neuroscience advancements via functional imaging studies have supported this view (Caplan, 2001; Sakai, 2005). In contrast, cognitive and construction linguistics have emphasized a single mechanism of both, yet neither theoretical framework has led to a computational theory (Perlovsky, 2009). Moreover, evolutionary linguistics has emphasized an evolving mechanism of language acquisition, yet results have led to incomputable complexities between language and cognition that cannot be disentangled (Perlovsky & Sakai, 2014). The interplay between language and cognition is one of the areas of focus in the subspecialty of *Cultural Neuropsychology,* defined as the systematic study of brain behavior relationships within the context of engagement in specific cultural practices that organize the development, maintenance, and revision of cognition/schemas and behaviors, revealing the relationship between sociocultural factors and cognition (Cagigas & Manly, 2014). This chapter will provide a brief historical overview of the influence of language on cognition, review terminology that is imperative to both clinical and research settings in the field of neuropsychology, introduce theoretical language theories, review possible

DOI: 10.4324/9781003051497-10

moderating and mediating factors in language development and mastery, and provide a synopsis of the documented disadvantages and advantages of bi/multilingualism on cognitive functioning via a lifespan perspective. The chapter will close with practical clinical considerations in the assessment of bi/multilingual individuals via an international lens.

Bilingualism was historically considered a handicap, limiting, or stunting cognitive abilities. In the United States, bilingualism was referred to as a disadvantage in reference to overall cognitive functioning – "This might be considered evidence that the use of a foreign language in the home is one of the chief factors in producing mental retardation as measured by intelligence tests" (Goodenough, 1926). As cross-cultural neuropsychologists, one can assume the many factors that were not considered to ensure culturally and linguistically appropriate assessment practices. Yet, it was not until 1962 that the first cognitive advantage secondary to bilingualism was documented, referred to as "mental flexibility" (Peal & Lambert, 1962). Fortunately, scientific gains have illustrated the disadvantages (mild and only present early in the lifespan) and advantages of language on cognitive functioning over time and as this line of research continues to evolve, consideration of confounding, mediating, and moderating factors is imperative. Recently included as a variable of importance in the study of language and cognition is culture/lifestyle practices, which impacts the development of language acquisition and cognitive schemas and provides a foundation, whether it be strong or weak, that guides skill development which may vary across languages, cognitive domains and other factors (i.e., environments, lifestyle practices, resources) (Boroditsky & Gaby, 2010).

Composed of different sounds, vocabulary, and structures (i.e., grammatical gender) language influences how we think/perceive stimuli, process information, and respond, not just verbally but in regard to reaction time. Cultural/lifestyle practices without consideration of language overlook cognitive processes intrinsic to culture. One example is the resulting differences between Western and Eastern regions regarding environmental focus and international education (Lun et al., 2010; Nisbett et al., 2001). Language affects cognitive processes in three basic ways. First, indirectly, by differences in where attention is directed in the environment (e.g., focus and priority on different parts of the situation/stimuli). Second, directly, by making some forms of social communication patterns more acceptable than others (e.g., usage of sustained eye contact during conversation varies by cultures). Third, by affecting the thinking style used to approach a novel situation (e.g., relying either heavily on critical thinking versus dialectical thinking). Thus, the reader is encouraged to review the following material via a cultural lens, with awareness of the interplay between culture, cognition and language (bi/multilingualism) that results in schema structures which impact cognitive outcomes (Ardila et al., 2017).

Terminology

The following terms are commonly utilized in the field of linguistics and are pertinent to understanding the influence of bi/multilingualism on cognitive performance. The definitions highlight linguistic details that must be considered and closely examined when working with a bi/multilingual individual (e.g., language types, age at acquisition, levels of language mastery, etc.).

Language types:

- Expressive: the ability to respond, request, and inquire. Non-fluent verbal skills include vocalizations and non-verbal variations comprise gesturing, writing, and facial expressions.
- Receptive: the ability to understand spoken or written information.
- Semantic: developing understanding and appropriate use of meaning not only of words, phrases, and sentences, but also the meaning of objects, concepts, and constructs. It is the ability to express oneself clearly and meaningfully with the understanding of the world.
- Pragmatic: social language skills that are evident in how messages are shared (i.e., tone and pace). Non-verbal output includes eye contact, facial expressions, and body language.

Basic interpersonal communication skills (BICS): language which is context embedded and facilitates communication in social environments and informal settings, such as socializing with peers and discussing day-to-day interactions. Abilities at this level can develop in two to three years (Cummins, 1984, 1999).

Cognitive academic language proficiency (CALP): language skills required in an academic setting, which include higher-order cognitive strategies. Language skills can be applied outside context areas and fully developed ability requires on average seven-ten years of academic exposure (Cummins, 1984, 1999).

Additive bilingualism/Balanced bilingualism: acquisition of two languages in a balanced manner; considered a strong level of bilingualism.

Early bilingualism: two types-simultaneous and consecutive/successive

- Simultaneous: a child who learns two languages at the same time, from birth. This generally produces a strong bilingualism, referred to as additive bilingualism. Language development is considered bilingual.
- Consecutive/Successive: a child who has partially acquired a first language and then learns a second language early in childhood. This generally produces a strong bilingualism (or additive bilingualism), but the child must be given time to learn the second language, as expressive and receptive skills are being learned simultaneously. Language development is considered partly bilingual.

Late bilingualism: bilingualism when the second language is learned after the age of six or seven; considered a consecutive bilingualism which occurs after the acquisition of the first language. With the development of the first language (L1 hereafter), the late bilingual uses their experience to learn the second language (L2 hereafter).

Semi-lingualism: a state of poor language proficiency in two or more languages (Cummins, 1981; Escamilla, 2006)

Subtractive bilingualism: acquisition of a second language to the detriment of the first language, especially if the first language is a minority language. In this case, mastery of the first language decreases, while mastery of the other language (usually the dominant language) increases.

Elite/Elected bilingual: acquisition of a second language via formal study.

Folk/Circumstantial bilingualism: native language is primarily oral, and its use is unsupported by formal education.

Passive/ Receptive bilingualism: able to understand a second language without being able to speak it.

Cross-cultural neuropsychology/cognitive neuroscience and language theories

The following section presents different models and hypotheses that posit how the brain adapts and operates when bi/multilingual individuals use more than one language to interact with their environment. These models outline the intersection of language, cognition, and brain functioning.

Competition and Entrenchment Hypothesis: age of exposure and proficiency impact language fluency (Kovelman et al., 2008). The competition and entrenchment model focuses on the process of learning a second language with brain plasticity and first-language entrenchment as factors contributing to differences in the fluency of L2 in comparison to L1. The *simultaneous* and early acquisition of L1 and L2 reduces the competition in linguistic processes because the entrenchment of L1 over L2 is minimal as each language develops a separate network. However, learning L2 *after* L1 has already become entrenched creates competition for L2 usage, as L2 is learned in relation to L1 without a separate network. This is to say that late bilinguals acquire their L2 throughout the lifespan with reduced neuroplasticity needing to activate more nonlinguistic processes, such as executive functioning, to improve L2 fluency (Hernandez et al., 2005).

Interference Hypothesis: cross-linguistic interference is common in bilinguals. When a word is activated in the target language, the same word is activated in the non-target language resulting in lexical access interference (Bijeljac-Babic et al., 1997; Duyck et al., 2007).

Inhibitory Control Model: this model presents a refinement of the interference hypothesis and suggests that individuals with bi/multi-languages must suppress irrelevant languages with the help of an inhibitory control mechanism. Thus, indicating that bilinguals must respond in one language, and suppress the other language whenever the presentation of two languages exists, known as global inhibition. Language use for bilinguals, therefore, involves a balance between interference and suppression. This balance permits the processing of stimuli and generating a fluent and appropriate response, known as local inhibition (Dijkstra & Van Heuven, 1998; Green, 1998; Guo et al., 2011).

Joint Activation Model: joint activation of languages creates an attention problem for bilinguals. When speaking, bilinguals select not only the structure, lexical choice, and syntax for a response, but also select the correct language. Therefore, these series of selected choices make linguistical processing effortful and costly for bilinguals (Bialystok et al., 2012).

In clinical application, these theories anchor our conceptualization when interpreting neuropsychological results and identifying patterns of performance. For example, a balanced bilingual patient may demonstrate strong executive functioning skills in working memory and novel problem solving as they learned L1 and L2 simultaneously (Competition and Entrenchment Hypothesis). On the other hand, verbal fluency skills may be relatively weaker because language selection and phonological/semantic choice are effortful and costly (Joint Activation Model).

Moderating/mediating factors in language development and cognitive performance

This section focuses on the most pertinent factors for consideration in the interpretation of a neurocognitive profile for a bi/multilingual examinee. Interpretation of data should be considered in conjunction with base rates and interpreted via a patient-centered approach in light of the possible moderating and mediating variables to ensure findings are representative of the individual's cognitive strengths and weaknesses. Disentangling the following is a complex task for the examiner and a clear and confirmatory conclusion may not always be possible, illustrating the limitations of our current measurement tools and our field in the assessment of culturally and linguistically diverse individuals.

General Cognitive Capacity: most pertinent in the early stages of life, an individual's general cognitive capacity must be considered when assessing language skills and the impact of bilingualism on cognitive functioning. Children with lower intelligence quotients may be slower at or unable to simultaneously master two languages and the influence of language on cognitive skill must be interpreted with general intellect as a foundation. Studies on intelligence reveal a strong relationship between intelligence and secondary language acquisition but only as far as academic skills are concerned. "The ability to perform well in standard intelligence tests correlates highly with school related second language learning, but is unrelated to the learning of a second language for informal and social functions" (Spolsky, 1989, p. 103; see Gardner, 1983 Multiple Intelligences).

Motivation: defined by Gardner and Lambert (1972), as an individual's overall goal or orientation (Ellis, 1985). Motivation should be assessed as acquisition or lack thereof may be confounded by this internal factor.

a. Integrative motivation: interest in the people and culture of the target language;
b. Instrumental motivation: language acquisition is functional.

Attitude: initially defined by Gardner and Lambert (1972) as persistence and further defined by Ellis, 1985 as the set of beliefs concerning the target language culture and the individuals to whom they are exposed, the individuals' own culture/identity, and the perception of the tasks/objectives involved. Feelings associated with the language should also be considered. Emotions may reflect other factors, such as linguistic difficulty/simplicity, difficulty/ease of learning, and/or social status (Richards, 1985).

Socioeconomic Status: children from below poverty levels have shown impoverished vocabulary in comparison to their counterparts with higher resources (Hart & Risley, 1995). These findings suggest that the addition of a second language may evidence a disadvantage in expressive and receptive language in the early stages of life.

Malnutrition/Nutrition: there is a growing body of literature suggesting a relationship between nutritional intake and neurocognitive development from pregnancy to childhood. In language development, higher maternal fish consumption has been associated with higher language and social skills in toddlerhood (Daniels et al., 2004). In contrast, mild but persistent malnutrition in early life (i.e., during the first two years of life) has been associated with lower cognition and language development, while supplementation with food can improve cognitive performance (Laus et al., 2011).

Parental Education: L1 and L2 development are transactional with cross-language influence (Paradis, 2011; Paradis et al., 2011). Parental language to the child varies (for example, vocabulary usage and amount of speech produced per day) according to the level of education. In turn, these differences in parental speech account for a large proportion of the differences in children's language ability (Hoff 2003; Pancsofar &Vernon-Feagans, 2006).

Migration: newly arrived families have less access to learning materials (Rodriguez et al., 2009). Factors linking migration and cognitive outcomes are multidimensional and complex and impact initial levels of cognitive function and the rate of cognitive changes over time (Xu et al., 2018).

Aptitude/Academic Exposure. Entering a school with low levels of literacy establishes an academic achievement trajectory (Campbell & Ramey, 1994). Effects of low print knowledge, phonological awareness, rapid serial naming, and verbal conceptual abilities at the beginning of kindergarten may persist in elementary grades, including fourth grade (Hecht et al., 2000).

Cognitive Reserve: the brain's capacity for functional compensation or resilience throughout aging or following damage (Cabeza & Dennis, 2013; Stern, 2012). In healthy aging, cognitive reserve refers to the relationship between the degree of brain damage/pathology and the intensity or onset of clinical manifestations (Stern, 2009).

Bilingual Cognitive Reserve: first reported by Bialystok et al. (2007), bilingualism is considered a protective factor in the onset of neurodegenerative conditions, as bilinguals are slower to exhibit signs of dementia. Studies have controlled for potential confounds (see Freedman et al., 2014 for a review).

Variables that impact language development and performance across cognitive domains should be considered in the assessment of bi/multilingual individuals. In addition to those presented, clinicians and researchers should be aware of psychosocial variables that may impact an individual's performance such as acculturation stressors, implicit bias, and stereotype threat (Clinton & Olvera, 2014), in addition to the influence of mood/mental health symptomatology.

Traditionally, pediatric neuropsychology has placed a greater emphasis on these variables in both clinical and research endeavors. Consideration of the aforementioned in adult and geriatric evaluations is critical and likely to yield benefit in the interpretation of an individual's cognitive profile across the adulthood.

Cognitive networks and neural correlates of cognitive reorganization across the lifespan

Pediatric Considerations: degree of neural changes is dependent on the age of acquisition and L2 proficiency level (Klein et al., 2006; Perani et al., 1998). Bi/multilingualism assessment in the pediatric stages of development may be considered a disadvantage in comparison to adulthood and geriatric phases of life. Verbal skills of bi/multilinguals in each language are generally weaker in comparison to their monolingual counterparts (Bialystok, 2017) and bi/multilingual individuals usually control a smaller vocabulary in each language (Bialystok et al., 2010). Performance on verbal fluency tasks reveal deficits for bilinguals, particularly in semantic fluency conditions, even if responses can be provided in either language (Gollan & Ferreira, 2009) and bilinguals are slower and less accurate than monolinguals on measures of confrontation naming (Hernandez et al., 2000; Ivanova & Costa, 2008). While disadvantages in the language domain may be more evident during the pediatric stage, research has shown that bilinguals of all ages demonstrate better executive control than monolinguals matched by age and other background factors (Bialystok, 2017). In children, executive control is central to academic achievement, and in turn, academic success is a significant predictor of long-term health and well-being (Duncan, 2010). In a meta-analysis, Adesope and colleagues (2010) calculated medium to large effect sizes for the executive control advantages in bilingual children.

Adult Considerations. While the bilingual advantage may be clearly identified during the pediatric stages of development, it is not always evident in samples of young adults. Across the lifespan, there is performance variability on behavioral measures of interference/conflict resolution. For example, the Simon task is an exercise that reveals a difference in accuracy/reaction time between trials in which stimulus and response are on the same side versus trials in which they are in opposite locations. When young children, young, middle-aged, and older adults completed the Simon tasks, results indicated a bilingual advantage in response time (RT) at both ends of the age spectrum (five-year-olds and older adults) and not in the young adult group (Bialystok et al., 2005). However, further research revealed that bilingual young adults outperformed their monolingual counterparts on the directional arrow Simon task, but only on the condition that included more monitoring and switching in comparison to the simple condition. Thus, the advantage appears to emerge when tasks/conditions increase in complexity (Bialystok et al., 2006). In sum, research has documented the bilingual advantage in the domain of executive control across the lifespan (Bialystok, 2017) and as previously noted, this skill is indirectly considered a significant predictor of long-term health and well-being (Duncan, 2010).

Epidemiological evidence suggests that older adults who engage in a multi-domain active lifestyle (e.g., incorporating social, cognitive, and physical activities) are protected to some degree against the onset of neurodegenerative conditions. Such factors are said to contribute to cognitive reserve, understood as the concept proposed to account for the disjunction between the degree of brain damage or pathology and its clinical manifestations (Stern, 2009). The bilingual advantage has been shown to extend into older age and protect against neurodegenerative processes (Adesope et al., 2010; Bialystok, 2017; Bialystok et al., 2007; Hilchey & Klein, 2011). Bialystok et al. (2007) were the first to present evidence for bilingual cognitive reserve as bilinguals showed initial signs of dementia four years later and were diagnosed 3.2 years later in comparison to their monolingual counterparts after controlling for gender, education, and employment status. Moreover, a more recent meta-analysis found that on average bilinguals exhibit symptoms of Alzheimer's disease for 4.1 years and are diagnosed with AD 2.0 years later than monolingual's (Paulavicius, 2020). Research has also confirmed a 4.7-year delay in AD onset and a 3.3-year delay in the diagnosis of dementia, but no delay in the diagnosis of mild cognitive impairment (Brini et al., 2020). Researchers have replicated these results, finding delays in the onset of neurodegenerative disease and examination of possible confounding variables have been extended to include depression, socioeconomic status diet, smoking, alcohol consumption, and physical and social activities (Craik, et al., 2010; Fergus et al., 2010; Freedman et al., 2014; Woumans et al., 2017).

Guzman-Velez and Tranel's work (2015) discussed potential neural mechanisms behind the proposed relationship and their conclusion indicated that lifelong bilingualism is related to more efficient use of brain resources that assist in maintaining cognitive functioning in the presence of neuropathology. Support for bilingualism as a protective factor against neurodegenerative conditions, specifically AD has continued. Anderson and colleagues (2020) revealed a moderate effect size for the protective effect of bilingualism on the age of onset of AD symptoms and weaker evidence that bilingualism prevents the occurrence of disease incidence itself (Cohen's $d = 0.10$). This was after accounting for the influence of education, SES, and publication bias. Overall, this line of research has revealed consistent delays related to bilingualism and onset of symptoms and diagnosis of AD and other dementias and has been used to highlight the phenomenon of *bilingual cognitive reserve*, and how a second language compensates for the effects of accumulated neuropathology.

The possible relationship between life-long bi/multilingualism and cognitive reserve is not fully supported. Several studies have failed to replicate the aforementioned (Clare et al., 2014; Crane et al., 2010; Lawton et al., 2015; Zahodne et al., 2014) and methodological caveats have been highlighted by Calvo et al. (2016), questioning the influence of bilingualism on neuroplasticity and cognitive reserve. These limitations include: 1. Shortcomings in the conception and assessment of language (bi/multilingualism); 2. Variability in sample design; 3. Reservations on the instruments used to assess and diagnose the underlying clinical entity; 4. Consideration of possible confounding variables (Calvo et al., 2016); and 5. Lack of information, including incidence rates and age or duration of symptom onset (Anderson et al., 2020).

Considering this work, the influence of life-long bi/multilingualism on cognitive reserve at the closing end of the lifespan has not been determined. In an attempt to better understand the possible bi/multilingual advantage in the later stages of life and its protective factor against neurodegenerative conditions, researchers are encouraged to explicitly compare opposing results, identify and examine possible confounding variables, understand the methodological frameworks used in past work and improve on operationalizing variables and methodological frameworks in future studies, carefully examine sample characteristics, and better understand and improve upon the measurement of language (bi/multilingualism).

Clinical implications and closing remarks

The neuropsychological assessment of a bi/multilingual individual is multifaceted and tailoring of the neuropsychological assessment begins with gathering information about the individual's background in order to inform test and normative data selection. The objective of such assessment is to provide a response to the referral question(s) while considering the engagement level (motivation, intellect, and linguistic skills) of the individual and the context in which the interpretation of neuropsychological data will be embedded. Neuropsychologists serving bi/multilingual individuals must be equipped with a strong foundation regarding the influence of sociodemographic and mediating/moderating psychosocial and environmental factors on standardized outcome measures. Careful consideration of the aforementioned variables (via structured and semi-structured practices in an attempt to reduce unconscious bias and stereotype threat) assists the examiner in reaching a better understanding of the neurocognitive profile of the examinee. In some cases, clear delineation of findings may not be possible due to confounding variables that cannot be disentangled. Results may not be clear suggesting an accumulation of negative influences on standardized testing practices (e.g., impoverished environment, low level of education, test naivete, disease, etc.). In such cases, neuropsychologists are encouraged to seek professional consultation with other examiners and/or cultural brokers that may assist in disentangling possible confounders. Further, the limitations of the young field of cultural neuropsychology must also be deliberated. It is with a clear understanding of the limitations of our measurements and normative data sets that we can work toward making improvements in our field in an attempt to better serve our quickly growing diverse communities.

Clinical Considerations

The following brief clinical guide provides suggestions for the preparation of a neurocognitive assessment with bilingual/multilingual individuals.

- Examiners strive to establish a clinical practice model that is welcoming and culturally informed to increase healthcare equity. For example, referral forms thoughtfully ask about sociocultural information and language usage of the examinee.
- If the examiner is unfamiliar with the cultural background and/or languages spoken by the bi/multilingual person, research to build cultural knowledge is necessary.
- Given the relation between cognition and language, it is of great importance to integrate questions about language acquisition and usage, educational history, and level of language skills/mastery in each language during the clinical interview.
- Examinees may indicate that one language is stronger than another, which may not be consistent with language testing results. Therefore, administration of language proficiency measures is ethically necessary to inform battery selection prior to beginning a comprehensive assessment.
- The use of medically certified interpreters is required when the neuropsychologist cannot directly communicate with the examinee.
- Measures of acculturation may be considered to gather information on language use across activities. A detailed language use interview, examining preferred languages across environments coupled with testing in each language is necessary to determine whether language difficulties reflect impairment or familiarity/fluency.
- Consider the examinee's point of view regarding their preferred testing language in reference to their neuropsychological visit. Clinical experience has revealed the examinee's concern regarding service quality, inferring a lower quality of care for non-English speakers. Patients have also reported feeling shy or embarrassed at their request of non-English services given experiences of racism and discrimination. The addition of an interpreter requires consent from the patient and providing this service before the day of the appointment will aid rapport. Review of confidentiality/HIPAA standards is imperative. Only medically certified interpreters should be utilized for this work. Interpreters must also be prepared/instructed on how to engage in neuropsychological practices. This is the responsibility of the examiner.
- Language block testing (testing on separate days/testing blocks) may be required to better understand language mastery and possible language impairment. Practice effects should be considered if testing blocks are used and sociodemographics of normative data across measures should be noted as this may impact normative cut-scores/clinical classifications.
- Documentation of normative data sets used for interpretation should be clearly stated in the neuropsychological report. This practice provides a reference not only to the reader of the report, but also for comparison should future assessments be needed.

- Confounding variables considered in the interpretation of findings should be noted by the examiner in the summary/impression portion of the report. This will assist the referral source and patient/family in understanding why at times clear distinctions in the neurocognitive profile (e.g., weaknesses versus impairment) may not be possible.
- Some testing practices or lack thereof, may be observed in bi/multilingual individuals (e.g., in the Latin American culture, there is a concern with accuracy at the cost of efficiency/speed; "despacio porque voy de prisa" Gilbert, 1986). Thus, mode of expression not only linguistically but behaviorally, must be considered, documented, and utilized in the interpretation of neuropsychological data.

Although significant scientific advancements have been made and our understanding of language and cognition continues to improve, examiners should be cautious in the interpretation of neurocognitive outcome measures of bi/multilingual individuals and those of non-Westernized cultural backgrounds to ensure that outcomes are interpreted in context and limitations of standardized tools and normative data sets are considered (refer to aforementioned clinical considerations). The introduction to this chapter highlights the complexity in disentangling culture from linguistic practices and we close this chapter in the same state, as the reviewed literature and conducted research aid in further highlighting this complexity for which we are currently unable to clearly measure and interpret. The limitation of our field should not be considered a barrier in the assessment of diverse populations as we need to focus on the need for ethical and equitable care. Concluding with caveats and additional questions can thus at times be the most appropriate conclusion, highlighting the need for additional examination (which may not be possible with our current tools), or data gathering via behavioral observations and collateral informants. Follow-up examination post interim interventions to address cognitive weaknesses or impairments may also be considered. If all domains were examined accordingly and findings were interpreted in context with language development, mediating/moderating factors, and medical and mental health variables and base rates, then the limitations noted are reflective of the status of our field and not the competency of the examiner. Noteworthy, the steps covered in reaching this impression should be clearly documented in the assessment process. Neuropsychology as a field has an urgent call to action to develop, adapt, and modify our current assessment tools and practices to evaluate diverse populations internationally, ethically, and equitably. In the interim, careful documentation of the field's limitations will aid patients/families and referral sources in better understanding the complexity of the evaluation process.

References

Adesope, O.O., Lavin, T., Thompson, T., & Ungerleider, C. (2010). A systematic review and meta-analysis of the cognitive correlates of bilingualism. *Review of Educational Research, 80*(2), 207–245.

Anderson, J.A., Hawrylewicz, K., & Grundy, J.G. (2020). Does bilingualism protect against dementia? A meta-analysis. *Psychonomic Bulletin & Review, 27,* 952–965.

Ardila, A., Cieślicka, A.B., Heredia, R.R., & Rosselli, M. (2017). *Psychology of bilingualism: The cognitive and emotional world of bilinguals.* Springer.

Bak, T.H., Nissan, J.J., Allerhand, M.M., & Deary, I. (2014). Does bilingualism influence cognitive aging? *Annals of Neurology, 75,* 959–963. doi: 10.1002/ana.24158.

Banham, V. (2014). Language: an important social and cultural marker of identity. Paper presented at the Language and Social Justice Issues Conference, held on 26 November 2014, at Edith Conway University, Perth, Australia.

Berwick, R. C., Friederici, A. D., Chomsky, N., & Bolhuis, J. J. (2013). Evolution, brain, and the nature of language. *Trends in Cognitive Sciences, 17*(2), 89–98.

Bialystok, E. (2017). The bilingual adaptation: How minds accommodate experience. *Psychological Bulletin, 143*(3), 233.

Bialystok, E., & Craik, F.I. (2010). Cognitive and linguistic processing in the bilingual mind. *Current Directions in Psychological Science, 19*(1), 19–23.

Bialystok, E., Craik, F.I., & Freedman, M. (2007). Bilingualism as a protection against the onset of symptoms of dementia. *Neuropsychologia, 45*(2), 459–464.

Bialystok, E., Craik, F.I., & Luk, G. (2012). Bilingualism: Consequences for mind and brain. *Trends in Cognitive Sciences, 16*(4), 240–250.

Bialystok, E., Craik, F.I., & Ryan, J. (2006). Executive control in a modified antisaccade task: Effects of aging and bilingualism. *Journal of Experimental Psychology: Learning, Memory, and Cognition, 32*(6), 1341.

Bialystok, E., Luk, G., Peets, K.F., & Yang, S. (2010). Receptive vocabulary differences in monolingual and bilingual children. *Bilingualism (Cambridge, England), 13*(4), 525–531. doi: 10.1017/S1366728909990423 [doi]

Bialystok, E., McBride-Chang, C., & Luk, G. (2005). Bilingualism, language proficiency, and learning to read in two writing systems. *Journal of Educational Psychology, 97*(4), 580.

Bijeljac-Babic, R., Biardeau, A., & Grainger, J. (1997). Masked orthographic priming in bilingual word recognition. *Memory & Cognition, 25*(4), 447–457.

Boroditsky, L., & Gaby, A. (2010). Remembrances of times east: Absolute spatial representations of time in an australian aboriginal community. *Psychological Science, 21*(11), 1635–1639.

Boroditsky, L. (2017, November). How language shapes the way you think [Video] TED. https://www.ted.com/talks/lera_boroditsky_how_language_shapes_the_way_we_think?language=en

Brini S., Sohrabi H.R., Hebert J.J., Forrest M.R.L., Laine M., Hämäläinen H., Karrasch M., Peiffer J.J., Martins R.N., Fairchild T.J. (2020, March) Bilingualism is associated with a delayed onset of dementia but not with a lower risk of developing it: A systematic review with meta-analyses. *Neuropsychol Rev. 30*(1):1–24. doi: 10.1007/s11065-020-09426-8. Epub 2020 Feb 8. Erratum in: Neuropsychol Rev. 2020 Mar 13;: PMID: 32036490; PMCID: PMC7089902. Cagigas, X.E., & Manly, J.J. (2014). *Cultural neuropsychology: The new norm.* In M.W. Parsons, T.A. Hammeke, & P.J. Snyder (Eds.), *Clinical neuropsychology: A pocket handbook for assessment* (pp. 132–156). American Psychological Association.

Calvo N., García A.M., Manoiloff L., & Ibáñez A. (2016) Bilingualism and cognitive reserve: A critical overview and a plea for methodological innovations. *Frontiers in Aging Neuroscience, 7:*249. doi: 10.3389/fnagi.2015.00249.

Campbell, F.A., & Ramey, C.T. (1994). Effects of early intervention on intellectual and academic achievement: A follow-up study of children from low-income families. *Child Development, 65*(2), 684–698.

Caplan, D. (2001). Functional neuroimaging studies of syntactic processing. *Journal of Psycholinguistic Research, 30*(3), 297–320.

Cagigas, X.E., & Manly, J.J. (2014). Cultural neuropsychology: The new norm. In M.W. Parsons, T.A. Hammeke, & P.J. Snyder (Eds.), *Clinical neuropsychology: A pocket handbook for assessment* (pp. 132–156). American Psychological Association. 10.1037/14339-008.

Clare, L., Whitaker, C.J., Craik, F.I., Bialystok, E., Martyr, A., Martin-Forbes, P., et al. (2014). Bilingualism, executive control, and age at diagnosis among people with early-stage Alzheimer's disease in Wales. *Journal of Neuropsychology.* doi: 10.1111/jnp.12061.

Clinton, A.B., & Olvera, P. (2014). Norm-referenced assessment and bilingual populations. *Academic Assessment and Intervention,* 102–113.

Craik, F.L., Bialystok, E., & Freedman, M. (2010). Delaying the onset of Alzheimer's disease: Bilingualism as a form of cognitive reserve. *Neurology 75,* 1726–1729.

Crane P.K., Gruhl, J.C., Erosheva, E.A., Gibbons, L.E., McCurry, S.M., Rhoads, K., et al. (2010). Use of spoken and written Japanese did not protect Japanese-American men from cognitive decline in late life. *Journals of Gerontology, Series B: Psychological Sciences and Social Sciences, 65,* 654–666. doi: 10.1093/geronb/gbq046.

Cummins, J. (1984). *Bilingualism and special education: Issues in assessment and pedagogy.* Clevedon, England: Multilingual Matters.

Cummins, J. (1999). *BICS and CALP: Clarifying the distinction* (ERIC Document Reproduction Service No. ED438551).

Daniels, J.L., Longnecker, M.P., Rowland, A.S., Golding, J., & ALSPAC Study Team-University of Bristol Institute of Child Health. (2004). Fish intake during pregnancy and early cognitive development of offspring. *Epidemiology,* 394–402.

Dijkstra, T., & Van Heuven, W.J. (1998). The BIA model and bilingual word recognition. *Localist Connectionist Approaches to Human Cognition,* 189–225.

Duncan, G. J., Ziol-Guest, K. M. , & Kalil, A. (2010). Early-childhood poverty and adult attainment, behavior, and health. *Child Development, 81*(1), 306–325.

Duyck, W., Assche, E.V., Drieghe, D., & Hartsuiker, R.J. (2007). Visual word recognition by bilinguals in a sentence context: Evidence for non-selective lexical access. *Journal of Experimental Psychology: Learning, Memory, and Cognition, 33*(4), 663–679. doi: 10.1037/0278-7393.33.4.663.

Ellis, R. (1985). A variable competence model of second language acquisition. *IRAL, 23*(1), 47–59.

Fergus, I.M.C., Bialystok, E., & Freedman, M. (2010). Delaying the onset of Alzheimer's disease: Bilingualism as a form of cognitive reserve. *Neurology, 75* (19).

Freedman, M., Alladi, S., Chertkow, H., Bialystok, E., Craik, F.I.M., Phillips, N.A., et al., (2014). Delaying the onset of dementia: Are two languages enough? *Behavioral Neurology, 2014,* 808137.

Gardner, R.C. (1983). Learning another language: A true social psychological experiment. *Journal of Language and Social Psychology, 2*(2–3–4), 219–239.

Gardner, R. C. , & Lambert, W. E. (1972). Attitudes and motivation in second-language learning.

Gelman, S., & Roberts, S. (2017). How language shapes the cultural inheritance of categories. *Proceedings of the National Academy of Sciences, 114*(30), 7900–7907. 10.1073/pnas.1621073114.

Gollan, T.H., & Ferreira, V.S. (2009). Should I stay or should I switch? A cost–benefit analysis of voluntary language switching in young and aging bilinguals. *Journal of Experimental Psychology: Learning, Memory, and Cognition, 35*(3), 640.

Goodenough, F.L. (1926). A new approach to the measurement of the intelligence of young children. *The Pedagogical Seminary and Journal of Genetic Psychology, 33*(2), 185–211.

Green, D.W. (1998). Mental control of the bilingual lexico-semantic system. *Bilingualism: Language and Cognition, 1*(2), 67–81.

Guo, T., Liu, H., Misra, M., & Kroll, J.F. (2011). Local and global inhibition in bilingual word production: FMRI evidence from Chinese–English bilinguals. *Neuroimage, 56*(4), 2300–2309.

Guzman-Velez, E. & Tranel, D. (2015). Does bilingualism contribute to cognitive reserve? Cogntiive and neural perspectives. *Neuropsychology, 29*(1), 139–150. 10.1037/neu0000105.

Hart, B., & Risley, T.R. (1995). *Meaningful differences in the everyday experience of young american children.* Paul H Brookes Publishing.

Hecht, S.A., Burgess, S.R., Torgesen, J.K., Wagner, R.K., & Rashotte, C.A. (2000). Explaining social class differences in growth of reading skills from beginning kindergarten through fourth-grade: The role of phonological awareness, rate of access, and print knowledge. *Reading and Writing: An Interdisciplinary Journal, 12*, 99–127.

Hernandez, A. E., Martinez, A., & Kohnert, K. (2000). In search of the language switch: An fMRI study of picture naming in Spanish–English bilinguals. *Brain and Language, 73*(3), 421–431.

Hernandez, A., Li, P., & MacWhinney, B. (2005). The emergence of competing modules in bilingualism. *Trends in Cognitive Sciences, 9*(5), 220–225.

Hilchey, M.D., & Klein, R.M. (2011). Are there bilingual advantages on nonlinguistic interference tasks? implications for the plasticity of executive control processes. *Psychonomic Bulletin & Review, 18*(4), 625–658.

Hoff, E. (2003). The specificity of environmental influence: socioeconomic status affects early vocabulary development via maternal speech. *Child Development, 74*, 1368–1378.

Holistic versus analytic cognition. *Psychological Review, 108*(2), 291–310. https://www.ncbi.nlm.nih.gov/pmc/articles/PMC5189637/

Ivanova, I., & Costa, A. (2008). Does bilingualism hamper lexical access in speech production? *Acta Psychologica, 127*(2), 277–288.

Kavé, G., Eyal, N., Shorek, A., & Cohen-Mansfield, J. (2008). Multilingualism and cognitive state in the oldest old. *Psychology and Aging, 23*, 70–78. doi: 10.1037/0882-7974 .23.1.70.

Klein, D., Zatorre, R.J., Chen, J., Milner, B., Crane, J., Belin, P., et al. (2006). Bilingual brain organization: A functional magnetic resonance adaptation study. *Neuroimage, 31*(1), 366–375.

Kovelman, I., Baker, S.A., & Petitto, L. (2008). Bilingual and monolingual brains compared: A functional magnetic resonance imaging investigation of syntactic processing and a possible "neural signature" of bilingualism. *Journal of Cognitive Neuroscience, 20*(1), 153–169.

Laus, M.F., Duarte Manhas Ferreira Vales, L., Braga Costa, T.M., & Sousa Almeida, S. (2011). Early postnatal protein-calorie malnutrition and cognition: a review of human and animal studies. *International Journal of Environmental Research and Public Health, 8*(2), 590–612.

Lawton, D.M., Gasquoine, P.G., & Weimer, A.A. (2015). Age of dementia diagnosis in community dwelling bilingual and monolingual Hispanic Americans. *Cortex*, *66*, 141–145. doi: 10.1016/j.cortex.2014.11.017.

Lun, V.M.C., Fischer, R., & Ward, C. (2010). Exploring cultural differences in critical thinking: Is it about my thinking style or the language I speak?. *Learning and Individual Differences*, *20*(6), 604–616.

Marian, V., & Spivey, M. (2003). Competing activation in bilingual language processing: Within-and between-language competition. *Bilingualism*, *6*(2), 97.

Miranda, C., Arce Rentería, M., Fuentes, A., Coulehan, K., Arentoft, A., Byrd, D., … & Rivera Mindt, M. (2016). The relative utility of three English language dominance measures in predicting the neuropsychological performance of HIV+ bilingual Latino/a adults. *The Clinical Neuropsychologist*, *30*(2), 185–200.

Nisbett, R.E., Peng, K., Choi, I., & Norenzayan, A. (2001). *Culture and systems of thought*: holistic versus analytic cognition. *Psychological review*, *108*(2), 291.

Pancsofar, N., & Vernon-Feagans, L. (2006). Mother and father language input to young children: contributions to later language development. *Journal of Applied Developmental Psychology*, *27*, 571–587.

Paradis, J. (2011). Individual differences in child english second language acquisition: Comparing child-internal and child-external factors. *Linguistic Approaches to Bilingualism*, *1*(3), 213–237.

Paradis, J., Genesee, F., & Crago, M.B. (2011). *Dual language development and disorders: A handbook on bilingualism and second language learning*. ERIC.

Paulavicius, A. M., Mizzaci, C. C., Tavares, D. R., Rocha, A. P., Civile, V. T., Schultz, R. R. , … & Trevisani, V. F. (2020). Bilingualism for delaying the onset of Alzheimer's disease: a systematic review and meta-analysis. *European Geriatric Medicine*, 1–8.

Peal, E., & Lambert, W.E. (1962). The relation of bilingualism to intelligence. *Psychological Monographs: General and Applied*, *76*(27), 1.

Perani, D., Paulesu, E., Galles, N.S., Dupoux, E., Dehaene, S., Bettinardi, V., et al. (1998). The bilingual brain. proficiency and age of acquisition of the second language. *Brain: A Journal of Neurology*, *121*(Pt 10), 1841–1852. doi:10.1093/brain/121.10.1841.

Perlovsky, L. (2009). Language and cognition. *Neural Networks*, *22*(3), 247–257.

Perlovsky, L., & Sakai, K. L. (2014). Language and cognition. *Frontiers in Behavioral Neuroscience*, *8*, 436.

Price, C.J. (2012). A review and synthesis of the first 20 years of PET and fMRI studies of heard speech, spoken language and reading. *Neuroimage*, *62*(2), 816–847.

Ransdell, S.E., & Fischler, I. (1987). Memory in a monolingual mode: When are bilinguals at a disadvantage? *Journal of Memory and Language*, *26*(4), 392–405.

Richards, J. C. (1985). *The context of language teaching*. (Vol. 3). Cambridge University Press.

Rodriguez, E.T., Tamis-LeMonda, C.S., Spellmann, M.E., Pan, B.A., Raikes, H., Lugo-Gil, J., et al. (2009). The formative role of home literacy experiences across the first three years of life in children from low-income families. *Journal of Applied Developmental Psychology*, *30*(6), 677–694.

Sakai, K. L. (2005). Language acquisition and brain development. *Science*, *310*(5749), 815–819.

Stern, Y. (2009). Cognitive reserve. *Neuropsychologia*, *47*(10), 2015–2028. 10.1016/j.neuropsychologia.2009.03.004.

Stern, Y. (2012). Cognitive reserve in ageing and Alzheimer's disease. *The Lancet Neurology*, *11*(11), 1006–1012.

Spolsky, B. (1989). Communicative competence, language proficiency, and beyond. *Applied Linguistics*, *10*(2), 138–156.

Suarez, P.A., Gollan, T.H., Heaton, R., Grant, I., Cherner, M., & HNRC Group (2014, March). Second-language fluency predicts native language stroop effects: evidence from Spanish-English bilinguals. *Journal of the International Neuropsychological Society*, *20*(3):342–348. doi: 10.1017/S1355617714000058. PMID: 24622502; PMCID: PMC4296729.

Weber-Fox, C., & Neville, H.J. (2001). *Sensitive periods differentiate processing of open-and closed-class words.*

Woumans, E., Versijpt, J., Sieben, A., Santens, P., Duyck, W. (2017). Bilingualism and cognitive decline: A story of pride and prejudice. *Journal of Alzheimer's Disease*, *60*(4):1237–1239. doi: 10.3233/JAD-170759. PMID: 28922163.

Xu, H., Vorderstrasse, A. A., McConnell, E. S., Dupre, M. E., Østbye, T., & Wu, B. (2018). Migration and cognitive function: a conceptual framework for Global Health Research. *Global Health Research and Policy*, *3*(1), 1–12.

Zahodne, L.B., Schofield, P.W., Farrell, M.T., Stern, Y., & Manly, J.J. (2014). Bilingualism does not alter cognitive decline or dementia risk among Spanish speaking immigrants. *Neuropsychology*, *28*, 238–246. doi: 10.1037/neu0000014.

9

THE INFLUENCE OF CULTURE ON THE ASSESSMENT OF EXECUTIVE FUNCTIONING

Jonathan Evans

Defining executive functions

Many definitions of executive functions have been suggested, making the construct somewhat difficult to define precisely (Jurado & Rosselli, 2007). Broadly, executive functions are the cognitive skills necessary to identify, work toward and achieve tasks or goals, and to modify our actions when required (Burgess & Simons, 2005). It has been argued that there are three core executive functions – inhibition, working memory and cognitive flexibility – which support "higher-order" processes such as reasoning, problem-solving, planning and task management (Diamond, 2013; Miyake et al., 2000)

Cultural influences on the development of executive functions

Executive functions develop throughout childhood and adolescence and are significant predictors of academic achievement and socio-emotional compe-tence (Schirmbeck et al., 2020). Several factors have been shown to be asso-ciated with the development of executive functions, including socio-economic status, bilingualism and parental scaffolding, and a number of studies have ex-amined the extent to which cultural context may influence the development of executive functions. Schirmbeck et al. (2020) conducted a systematic review of cross-national studies that have examined the development of executive func-tions in children from pre-school to adolescence. A total of 26 studies were included, with study samples from 28 countries. A large proportion of the studies compared children from Western and East Asian countries, but there was a range of other cross-national comparisons. Bilingual children outperformed monolingual peers, with performance benefits evident on tasks of shifting and

DOI: 10.4324/9781003051497-11

inhibition. Across the lifespan, it appears that bilingualism confers an advantage on some executive function tasks (e.g., tests of inhibition and verbal fluency) but only at time points when these skills are undergoing developmental changes (Zeng et al., 2019)

In the Schirmbeck et al. (2020) review, gender differences in executive functions varied between countries – in Euro-American and East Asian countries girls tend to outperform boys (with small effect sizes), though in Tanzania boys performed above the level of girls. On parental rating measures girls tended to be rated higher than boys in most countries (e.g., Sweden, Spain, China) but in Iran boys were rated higher than girls. Another general finding was that East Asian children tended to outperform their Western counterparts on direct tests of executive functions, particularly tests of inhibition. Results for tests of shifting were more mixed. The gap between East Asian and Western children is evident at a young age and widens in adolescence. The precise mechanisms that would explain the differences between cultures in the development of executive functions (or differences in performance on tasks considered to test executive functions) are not clear, though Schirmbeck et al. (2020) noted speculation that cultural differences in teachers' classroom management styles may influence the cognitive and behavioral self-regulation skills of children.

In most countries, executive functions increased with age, but in some studies in countries described as "severely underprivileged economically" older children did not consistently score above their younger peers. For example, a comparison of the US and South-African children found that there was the expected increase in cognitive flexibility skills in three-five-year-old American children on a task that involves switching between dimensions on a card sorting task, but this age-related change was not evident for the South-African children, even though they showed age-related changes in word learning (Legare et al., 2018). Legare et al. speculated that rule-switching flexibility may be dependent upon particular cultural experiences. They note that their US sample typically attended pre-schools in which they may have had experience of activities that impose explicit symbol-mappings, with participation in rule-based games. This raises an interesting issue of whether these sorts of educational activities train specific executive functions, or provide exposure to particular tasks that make demands on executive functions, and which confer an advantage on tasks with a similar format. In another study (Holding et al., 2018), children aged seven-18 years from Bangladesh, Ghana and Tanzania completed a range of cognitive tasks, including tests of executive function. On a test of shifting (changing responses to a series of hand movements when a specific trigger movement is displayed) children from Tanzania and Bangladesh showed age-related changes, but children from Ghana did not.

It is clear that there is a range of factors that influence the acquisition of executive functions, some of which may vary systematically between cultural contexts, and educational experience may particularly influence the development of cognitive skills to which common tests of executive functions are sensitive.

Assessment of executive functions

Ardila (1995) suggested that "comparable cognitive disturbances associated with brain pathology have similar manifestations across members of the human species" (p. 143). Thus, we might expect damage to the frontal lobes of the brain to have a broadly similar impact on executive functions whatever a person's cultural background. However, Ardila also noted that "cognitive abilities measured by neuropsychological tests represent, at least in their contents, culturally learned abilities" (p. 143). This is particularly the case for the assessment of executive functions.

Given that the complexity of the concept of executive functions it is not surprising that there is little agreement on how to assess executive functions. There are a number of tests that are widely cited as measures of various aspects of executive functions, many of which emerged as a means of detecting damage to the frontal lobes, and these will be considered first as they are also the executive tests that have been studied most frequently across cultures.

Perhaps the best-known test of inhibition is the Stroop test, or the Color-word interference test, originally developed by J.R. Stroop in the USA in 1935 (Stroop, 1935). The basic form of this task involves three sub-tasks – naming colored rectangles; reading color words written in black ink; naming the color of words that are printed in an incongruent color (e.g., the word blue written in red ink, for which the response required is red). Although the task is described as a test of inhibition (inhibiting the prepotent response of reading the word and instead of naming the color of the incongruent ink), it can be considered to make demands on selective attention (attending to the ink rather than the word), strategy application (e.g., attending to just one letter to avoid the temptation to read the word) and speed of information processing (as it is a timed task).

Performance on the Stroop test is associated with a number of demographic variables including age, sex, and education (Van der Elst et al., 2006). In a Dutch study, Van der Elst et al. (2006) found that educational level had a strong impact on Stroop performance, and also interacted with age – older people with low levels of education were particularly disadvantaged. While this study was not a cross-cultural study (all participants were Caucasian and native Dutch speakers), it highlights the impact of education on task performance, and education is a factor that varies considerably across countries.

Many Stroop studies have compared participants from different linguistic/ cultural contexts and reported differences between cultures on the critical "interference" trial. Alansari and Baroun (2004) note that studies have found greater interference for Chinese-speaking participants than English speaking, as well as for Chinese than Japanese, for Spanish than English, for Swedish than German, for German than English, and for English than French. In their own study, Alansari and Baroun compared Kuwaiti and British students, using Arabic and English versions of the test, and found that Kuwaiti students showed significantly slower performance on the interference trial than the British, suggesting Kuwaitis were more vulnerable to interference than the British. Alansari and Baroun speculate that different language structures of Arabic and English might explain the discrepancies, with Arabic being more vulnerable to interference because Arabic words often have at least two pronunciations, something that may create greater distraction/interference. They also raise the possibility that the collectivist/individualistic distinction that has been used to explain differences in attention to scene perception may be relevant to the Stroop task with people from collectivist cultures being more vulnerable to interference as they attend more to context than those from individualistic cultures. A more mundane explanation for Alansari and Baroun's findings, however, is that Arabic color words are typically longer than English color words and so this means they take longer to read, meaning that all conditions that require saying colors will be slower. This was evident as Kuwaitis were slower on the simple word reading or color naming conditions. This highlights how important it is to ensure that apparent cross-cultural differences are not emerging as a result of a failure to ensure tests are equivalent (albeit this can be challenging to determine). It also highlights that the executive component of the Stroop test should be examined by a comparison of performance in interference conditions that controls for performance in the simpler conditions.

What is evident from the Stroop literature is that performance differences on tests such as Stroop will exist between people from different cultural/linguistic backgrounds for a variety of reasons. This does not mean that the Stroop is not appropriate to use in different cultures but simply that its psychometric properties (reliability/validity) and normative data must be established in each new context in which it will be used and that each context must ensure that normative samples reflect the range of factors that are known to impact performance.

Another major class of executive function tests are tests of mental flexibility, typically examined by tasks that require switching between mental sets. Examples include sorting tasks such as the Wisconsin Card Sorting Test and variants of the Trail Making Test (TMT). The classic form of the TMT consists of two parts. In Part A, numbers are scattered around a sheet and the subject must connect numbers in ascending order. In Part B there are numbers and letters and the subject must connect alternatively numbers (in ascending order) and letters (in alphabetical order). Part A is considered to make demands on visual search, attention and speed of information processing, while Part B makes additional demands on mental flexibility, and the ability to switch between two mental sets. TMT is affected by age, education and IQ and therefore may be different across cultural contexts that differ systematically in access to education. As the test uses numbers and letters it may not be suitable for those with low levels of numeracy/literacy. Fernandez and Marcopulos (2008) examined normative data from 11 countries to examine potential sources of bias on test performance. Their key finding was that norms are not interchangeable, even between Western (or Westernized) countries – people who are clearly healthy in one context would be classified as impaired in another. An important source of bias was that samples across countries differed in terms of levels of education, occupation, and intelligence. Fernandez and Marcopulos (2008) also noted that there is evidence that TMT may not even assess the same construct across cultures. For example, Lee et al. (2000), compared participants from different cultures on the TMT and the Color Trails Test (CTT; designed to make similar demands to the TMT but without the requirement for knowledge of the alphabet), finding that English participants showed strong correlations between the TMT and CTT, whereas Chinese participants did not. Another study examined the performance of Farsi-speaking Iranians who had immigrated to the US on the WCST, TMT, and CTT (Avila et al., 2019). They found that a high proportion of their sample scored greater than one standard deviation below the mean of the US normative data – 78% for TMT B; 50% for CTT2l 79% for WCST categories completed. Avila et al. discuss the possibility that even if the level of education was not low (average education was 13 years), the nature of education may have been different, with US education more strongly emphasizing abstraction and problem-solving skills that are required for neuropsychological tests. But they also note that the TMT was adapted to use Farsi letters. Farsi is an orthographically complex language and rote memorization of the alphabet is not encouraged in the same way as it is in English, meaning that it may have simply taken longer to work out the next letter in Farsi. Nevertheless, differences were also apparent on the CTT, which should be less affected by linguistic differences.

Tests of working memory, such as backward digit span, have long been shown to be affected by education (Ostrosky-Solis & Lozano, 2006; Sierra Sanjurjo et al., 2019). Ostrosky-Solis and Lozano (2006) found that backward digit span was significantly affected by education in a Mexican sample. They note that formal education, particularly learning to read and write, "promotes an alternative way in which information can be conceptually processed" (p. 339) and it has been shown that people who are illiterate who are taught to read and write improve their neuropsychological test performance, including on digit span. Ostrosky-Solis and Lozano also reviewed studies that had attempted to examine the effects of age, education, and culture on digit span performance. They identified three studies (including participants from ten countries) and also compared their own data from Mexico with that in the published studies. Their approach was to match a subset of participants from their own sample with the mean age and education level of the sample they were comparing. What they found was that their Mexican sample scored significantly below that of almost all of the other samples despite being matched on length of education.

Ostrosky-Solis and Lozano note that their study suggests that there are differences in digit span performance between different samples that cannot be explained by just age or years of schooling, with the implication that there are additional "cultural factors" that must account for the differences. They discuss first the possibility that differences in speech rate between different languages may explain digit span differences, but note that this explanation has only been shown in relation to forward digit span and not backward digit span. Another explanation considered is that while there may not be differences in the quantity of schooling, there may be differences in the quality of education. The issue of quality of education is discussed in more detail later, but Ostrosky-Solis and Lozano suggest that differences between education systems such as when children learn to read and write may impact brain functional organization, and as a result impact performance on digit span tasks, though the precise relationship between aspects of education, and cognitive functions that support the ability to do backward digit span tasks is not clear.

Even if normative performance between cultures is different, this does not necessarily mean that the test will not be valid for the purpose it is intended. For example, verbal fluency tasks are another type of test that is considered to reflect both verbal and executive functions. Both letter and category fluency tasks are considered to reflect executive functions of updating and inhibition (Shao et al., 2014). A systematic review of neuropsychological tests for the detection of dementia in non-Western, low-educated, or illiterate populations (Franzen et al., 2020) reported that 12 studies had investigated the use of verbal fluency tasks (mostly category fluency- animals, fruit, vegetables) in the detection of dementia and found fair to excellent diagnostic accuracy, though noted that both sensitivity and specificity were lower for lower-educated participants. What is important here is that for some tests normative performance may differ between cultural contexts, but the test may still be valid for its intended purpose, such as contributing to diagnosis of dementia.

Universal assessment of executive functions?

"Cross-cultural" batteries of neuropsychological tasks have been developed with the explicit aim that the tests could be applied across ethnic groups and languages without the need to change the content and be applied with people who have little or no education. One example is the European Cross-Cultural Neuropsychological Test Battery (Nielsen et al., 2018). This battery includes tests of global cognitive function, memory, language, executive and visuospatial functions. The tests of executive function are: (i) the Color Trails Test – a non-alphabetical form of the Trail Making Test; (ii) The Five Digit Test – a form of Stroop test designed to minimize the need for reading ability that has three conditions including naming 50 digits, counting a series of asterisks and counting a series of digits in which the numeric value of the digits is incongruent with the number of digits (e.g., two 5's three 4's, five 1's); (iii) Serial threes – a simpler version of Serial Sevens, where participants count down from twenty by three. Nielsen et al. (2018) examined the performance of participants from a range of backgrounds including western European majority participants and people from various minority ethnic backgrounds (Moroccan; Pakistani/Indian Punjabi; Polish; Turkish; and former Yugoslavian). They found that there were significant differences between ethnic groups on most of the battery, including all three of the executive tasks. However, education explained the most variance in performance on executive tests between minority ethnic groups, explaining 10–33% of variance depending on the test. A much smaller proportion of variance (3–10%) was explained by ethnic group once education was controlled.

Nielsen et al. (2018) note that other attempts to create batteries suitable for people with limited education have not been able to eradicate the influence of education on test scores. They suggest that this can be explained by the fact that even though tests are designed not to require knowledge that would be acquired at school, the way that tests are performed will be affected by experience acquired in a school context. It is interesting that executive tests appear to be particularly vulnerable to this compared to some other domains. For example, Nielson et al. included the Recall of Pictures Test as one of their tests of memory, which is similar to a word list learning test but uses pictures of common objects. This test showed almost no influence on educational level. By contrast, animal fluency, Color Trails and the Five Digit Test were all strongly affected by education.

Quality of education and executive functions

Many normative studies of neuropsychological test performance include a measure of educational level, typically years of education. However, it is also apparent from several studies that years of education may not be the best measure of the impact of education on test performance. This is because quality of educational experience may vary widely. Measures of *quality* of education may be much better than measures of quantity. Manly et al. (2002) found that differences in performance on a range of cognitive tests between African Americans and non-Hispanic White Americans who were matched for years of education were greatly reduced when scores were adjusted for reading level (measured using the Reading Recognition subtest of the Wide Range Achievement Test – Version 3). Manly et al. suggested that years of education do not capture the educational experience among multi-cultural elders. There was limited assessment of executive functions in the Manly et al. (2002) study, but a similar study by Schneider and Lichtenberg (2011) found that reading ability, but not education, was significantly associated with performance on the Trail Making Test, Controlled Oral Word Association (COWA) Test, Animal Naming, Digit Span and Stroop Tests. With a specific focus on executive functions, Johnson et al. (2006) also found that reading ability accounted for a significantly greater amount of variance than years of education for Letter-Number Sequencing, Similarities, COWA, Trail Making Test, and Colored Progressive Matrices. Furthermore, reading ability mediated the relationship between the tests and level of education. Interestingly, category fluency was not influenced by educational quality, which is surprising given that it was in other studies, such as Nielsen et al. (2018). It is argued that a failure to consider education (and particularly quality of education) may contribute to the misclassification of minority populations.

What is the nature of the relationship between reading ability and performance on this wide range of executive function tests? It seems unlikely that it is actually reading ability that is critical as most of the executive tests do not involve much reading. One option is that reading ability is associated with other aspects of quality of education such as teacher-student ratios, access to learning resources, hours of teaching. Alternatively reading ability may be a proxy for socio-economic status which in turn relates to factors such as access to resources at home, diet, parental support, etc. Johnson et al. (2006) also noted that reading ability could be a correlate of g, or general intellectual ability, and it is variation in g that explains the differences on tests of executive function. But again, this leads to the question of what social/demographic/cultural factors are impacting development of the cognitive skills and knowledge required to do tests of intelligence used to measure g? Fagan (2000) draws the distinction between "processing" and knowledge, arguing that tests of intelligence require both processing ability and knowledge. Education provides exposure to information that provides knowledge which assists performance on tests of intelligence, including on many tasks that might be considered to be "culture-free". This is nicely illustrated in studies of children born just before or just after an arbitrary cut-off date for school entry that show that children born just before the cut off (and hence have more schooling) perform significantly better than those born just after the cut-off, including on tests such as Raven's Progressive Matrices.

Although the underlying mechanisms that account for cross-cultural differences in tests of executive function remain elusive, the clinical implications are clear. Any adaptation of an existing test of executive functioning for use in a new cultural and linguistic context requires the examination of the test's reliability and validity for its intended purpose in that context, as well as the collection of a context-specific normative sample (unless it can be demonstrated that the test performs in a very similar way to how it has performed in its original context). For example, the UK version of the WAIS IV was shown to perform in an equivalent way to the original US normative sample and hence UK psychologists use US norms to interpret test performance (Wechsler, 2008).

Ecologically valid tests of executive function

So far in this chapter, the focus has been on traditional tests of executive function. However, various other measures of executive functions have been developed that are considered to be more "ecologically valid" and more sensitive to deficits in cognitive skills such as planning, problem-solving, and task management. Shallice and Burgess (1991) developed two tasks, the Multiple Errands Test and the Six Elements Test that were shown to be more sensitive than traditional tests to everyday organizational problems arising following frontal lobe damage in three patients. These patients passed most of the traditional tests, but nevertheless had significant difficulties with the new, more ecologically valid tasks, that require planning, prospective memory, and task management over an extended period of time (10–15 mins). The Six Elements Test was subsequently modified and incorporated into a battery of executive function tasks, the Behavioral Assessment of the Dysexecutive Syndrome (BADS; Wilson et al., 1996). Chan and Manly (2002) compared performance of healthy Hong Kong Chinese participants with the original UK normative sample on the Modified Six Elements test and the Dysexecutive Questionnaire from the BADS and reported that performance was similar. Kallambettu et al. (2017) compared South Asian with White adults on the BADS and reported no significant differences in performance. Finally, Ihara et al. (2003) found that the BADS was sensitive to executive function impairment in participants with a diagnosis of Schizophrenia from both Japan and the UK, concluding that the study showed evidence of "culture-neutral neurobehavioral processes".

The Jansari Executive Function (JEF) assessment combines non-immersive virtual reality with paper & pencil tasks to assess a range of executive functions in an ecologically relevant context. The adult version (Denmark et al., 2019) involves a variety of tasks in an office environment while the children's version (Gilboa et al., 2019) involves organizing a children's birthday party. A recent study (Simon et al., 2020) adapted the test into Hebrew for use in Israel and reported that only relatively minor adaptations were required to ensure its cultural relevance.

Although cross-cultural research on these ecologically relevant tests of executive function is more limited than studies of more traditional tests, there are indications that the practical "everyday" tasks may work well across cultures. However, some caution is required as the few studies to date have primarily involved participant samples with good levels of education.

Conclusion

This chapter has highlighted that people from different linguistic/cultural contexts perform differently on traditional tests of executive function. Simple language differences always need to be considered in translations/adaptations of tests. There are multiple other influences on the development of the cognitive skills required to perform executive tasks, but bilingualism and educational experience are probably the most significant factors. Exactly what aspects of experience in education influence the development of executive functions is not clear, but it is likely that education exposes children, adolescents and adults to tasks and activities that provide knowledge required to undertake tests that make demands on executive functions. Hence differences in education will have a significant impact on test performance. Other cultural factors such as approach to timed tasks can also influence performance.

As with all domains of cognition, the implication of the findings discussed here is that the use of a test of executive function in a linguistic and cultural context that is different from the context in which the test was first developed requires careful consideration. Different linguistic/cultural contexts exist within countries and between countries. It is necessary to consider whether the constructs under investigation are relevant in the new context. Even if relevant, adaptation may be required and examination of reliability and validity will be necessary. New normative data are likely to be required, unless equivalence with the original sample is established.

References

Alansari, B., & Baroun, K. (2004). Gender and cultural performance differences on the Stroop Color and Word Test: A comparative study. *Social Behavior and Personality, 32*(3), 235–245. Retrieved from <Go to ISI>://WOS:000221575200004.

Ardila, A. (1995). Directions of research in cross-cultural neuropsychology. *Journal of Clinical and Experimental Neuropsychology, 17*(1), 143–150. doi:10.1080/13803399508406589.

Avila, J.F., Verney, S.P., Kauzor, K., Flowers, A., Mehradfar, M., & Razani, J. (2019). Normative data for Farsi-speaking Iranians in the United States on measures of executive functioning. *Applied Neuropsychology-Adult, 26*(3), 229–235. doi:10.1080/232 79095.2017.1392963.

Burgess, P., & Simons, J.S. (2005). Theories of frontal lobe executive function: Clinical applications. In P.W., Halligan , & D.T. Wade, (Eds.) *Effectiveness of rehabilitation for cognitive deficits.* (pp.211–231). Oxford: Oxford University Press.

Chan, R.C.K., & Manly, T. (2002). The application of "dysexecutive syndrome" measures across cultures: Performance and checklist assessment in neurologically healthy and traumatically brain-injured Hong Kong Chinese volunteers. *Journal of the International Neuropsychological Society, 8*(6), 771–780. doi:10.1017/s1355617702860052.

Denmark, T., Fish, J., Jansari, A., Tailor, J., Ashkan, K., & Morris, R. (2019). Using virtual reality to investigate multitasking ability in individuals with frontal lobe lesions. *Neuropsychological Rehabilitation, 29*(5), 767–788. doi:10.1080/09602011.2017.1330695.

Diamond, A. (2013). Executive functions. *Annual Review of Psychology, 64* (64), 135–168. doi:10.1146/annurev-psych-113011-143750.

Fagan, J.F. (2000). A theory of intelligence as processing – implications for society. *Psychology Public Policy and Law, 6*(1), 168–179. doi:10.1037//1076-8971.6.1.168.

Fernandez, A.L., & Marcopulos, B.A. (2008). A comparison of normative data for the Trail Making Test from several countries: Equivalence of norms and considerations for interpretation. *Scandinavian Journal of Psychology, 49*(3), 239–246. doi:10.1111/j.1467-945 0.2008.00637.x.

Franzen, S., van den Berg, E., Goudsmit, M., Jurgens, C.K., van de Wiel, L., Kalkisim, Y., ... Papma, J.M. (2020). A Systematic review of neuropsychological tests for the assessment of dementia in non-Western, low-educated or illiterate populations. *Journal of the International Neuropsychological Society, 26*(3), 331–351. doi:10.1017/s1355617719000894.

Gilboa, Y., Jansari, A., Kerrouche, B., Ucak, E., Tiberghien, A., Benkhaled, O., ... Chevignard, M. (2019). Assessment of executive functions in children and adolescents with acquired brain injury (ABI) using a novel complex multi-tasking computerised task: The Jansari assessment of Executive Functions for Children (JEF-C (c)). *Neuropsychological Rehabilitation, 29*(9), 1359–1382. doi:10.1080/09602011.2017.1411819.

Holding, P., Anum, A., van de Vijver, F.J.R., Vokhiwa, M., Bugase, N., Hossen, T., ... Gomes, M. (2018). Can we measure cognitive constructs consistently within and across cultures? Evidence from a test battery in Bangladesh, Ghana, and Tanzania. *Applied Neuropsychology: Child, 7*(1), 1–13. doi:10.1080/21622965.2016.1206823.

Ihara, H., Berrios, G.E., & McKenna, P.J. (2003). The association between negative and dysexecutive syndromes in schizophrenia: A cross-cultural study. *Behavioural Neurology, 14*(3-4), 63–74. doi:10.1155/2003/304095.

Johnson, A.S., Flicker, L.J., & Lichtenberg, P.A. (2006). Reading ability mediates the relationship between education and executive function tasks. *Journal of the International Neuropsychological Society, 12*(1), 64–71. doi:10.1017/s1355617706060073.

Jurado, M.B., & Rosselli, M. (2007). The elusive nature of executive functions: A review of our current understanding. *Neuropsychology Review, 17*(3), 213–233. doi:10.1007/s11 065-007-9040-z.

Kallambettu, V., Burda, A.N., & Wakeman, N. (2017). South Asian adults' performance on executive function tests. *American Journal of Speech-Language Pathology, 26*(4), 1254–1261. doi:10.1044/2017_ajslp-16-0173.

Lee, T.M.C., Cheung, C.C.Y., Chan, J.K.P., & Chan, C.C.H. (2000). Trail making across languages. *Journal of Clinical and Experimental Neuropsychology, 22*(6), 772–778. doi:10.1 076/jcen.22.6.772.954.

Legare, C.H., Dale, M.T., Kim, S.Y., & Deak, G.O. (2018). Cultural variation in cognitive flexibility reveals diversity in the development of executive functions. *Scientific Reports, 8*. doi:10.1038/s41598-018-34756-2.

Manly, J.J., Jacobs, D.M., Touradji, P., Small, S.A., & Stern, Y. (2002). Reading level attenuates differences in neuropsychological test performance between African American and White elders. *Journal of the International Neuropsychological Society, 8*(3), 341–348. doi:10.1017/s1355617702813157.

Miyake, A., & Friedman, N.P. (2012). The nature and organization of individual differences in executive functions: Four general conclusions. *Current Directions in Psychological Science, 21*(1), 8–14. doi:10.1177/0963721411429458.

Miyake, A., Friedman, N.P., Emerson, M.J., Witzki, A.H., Howerter, A., & Wager, T.D. (2000). The unity and diversity of executive functions and their contributions to complex "frontal lobe" tasks: A latent variable analysis. *Cognitive Psychology, 41*(1), 49–100. doi:10.1006/cogp.1999.0734.

Nielsen, T.R., Segers, K., Vanderaspoilden, V., Bekkhus-Wetterberg, P., Minthon, L., Pissiota, A., ... Waldemar, G. (2018). Performance of middle-aged and elderly European minority and majority populations on a Cross-Cultural Neuropsychological Test Battery (CNTB). *Clinical Neuropsychologist, 32*(8), 1411–1430. doi:10.1080/13854 046.2018.1430256.

Ostrosky-Solis, F., & Lozano, A. (2006). Digit span: Effect of education and culture. *International Journal of Psychology, 41*(5), 333–341. doi:10.1080/00207590500345724.

Schirmbeck, K., Rao, N., & Maehler, C. (2020). Similarities and differences across countries in the development of executive functions in children: A systematic review. *Infant and Child Development, 29*(1). doi:10.1002/icd.2164.

Schneider, B.C., & Lichtenberg, P.A. (2011). Influence of reading ability on neuropsychological performance in African American elders. *Archives of Clinical Neuropsychology, 26*(7), 624–631. doi:10.1093/arclin/acr062.

Shallice, T., & Burgess, P.W. (1991). Deficits in strategy application following frontal-lobe damage in man. *Brain, 114*, 727–741. doi:10.1093/brain/114.2.727.

Shao, Z., Janse, E., Visser, K., & Meyer, A.S. (2014). What do verbal fluency tasks measure? Predictors of verbal fluency performance in older adults. *Frontiers in psychology, 5*, 772. doi:10.3389/fpsyg.2014.00772.

Sierra Sanjurjo, N., Belen Saraniti, A., Gleichgerrcht, E., Roca, M., Manes, F., & Torralva, T. (2019). The IFS (INECO Frontal Screening) and level of education: Normative data. *Applied Neuropsychology-Adult, 26*(4), 331–339. doi:10.1080/23279095.2018.1427096.

Simon, N.O., Jansari, A., & Gilboa, Y. (2020). Hebrew version of the Jansari assessment of Executive Functions for Children (JEF-C (c)): Translation, adaptation and validation. *Neuropsychological Rehabilitation.* doi:10.1080/09602011.2020.1821718.

Stroop, J.R. (1935). Studies of interference in serial verbal reactions. *Journal of Experimental Psychology, 18*, 643–662. doi:10.1037/h0054651.

Stuss, D.T. (2011). Functions of the frontal lobes: Relation to executive functions. *Journal of the International Neuropsychological Society, 17*(5), 759–765. doi:10.1017/s1355617711 000695.

Van der Elst, W., Van Boxtel, M.P.J., Van Breukelen, G.J.P., & Jolles, J. (2006). The stroop color-word test − Influence of age, sex, and education; and normative data for a large sample across the adult age range. *Assessment, 13*(1), 62–79. doi:10.1177/10731911052 83427.

Wechsler, D. (2008). *Wechsler Adult Intelligence Scale, Fourth UK Edition Administration and Scoring Manual.* London: Pearson.

Wilson, B.A., Alderman, N., Burgess, P., Emslie, H., & Evans, J.J. (1996). *The behavioural assessment of the dysexecutive syndrome.* London: Pearson.

Zeng, Z., Kalashnikova, M., & Antoniou, M. (2019). Integrating bilingualism, verbal fluency, and executive functioning across the lifespan. *Journal of Cognition and Development, 20*(5), 656–679. doi:10.1080/15248372.2019.1648267.

PART III

Practical consequences of the influence of culture on clinical practice in neuropsychology

10

CROSS-CULTURAL TESTING: ADAPTATION, DEVELOPMENT, OR CROSS-CULTURAL TESTS?

Alberto Luis Fernández and Jonathan Evans

Introduction

The United Nations estimated in 2019 that there were 271,642,105 immigrants (someone who comes to live in a country from another country) in the world (United Nations, 2019), all of whom need access to health services and some of whom will require neuropsychological services. Neuropsychologists across the world are therefore faced with situations in which they must assess a person who speaks a different language and whose cultural experiences may be different from their own. A key feature of neuropsychological assessment is the administration of cognitive tests. These tools allow the clinician to obtain information in an objective, valid, reliable, quick, and economic way, as compared to using alternative assessment methods (interviews, observations, etcetera). Numerous articles describe the cultural bias of many of the most frequently used neuropsychological tests (Ardila, 2007; Fernandez & Abe, 2018; Nell, 2000). A test developed in one culture might not be appropriate for individuals raised in a different culture. Furthermore, inappropriate use might lead to incorrect diagnosis, with significant consequences for the patient. In order to properly assess a patient from a different cultural background, it is important to use tests that are not culturally biased or are appropriate for the culture of this specific patient.

There seem to be three major options to deal with this situation: adapting currently existing tests, developing specific tests for each culture or developing cross-cultural tests (CCTs; tests that are shown to be appropriate in many or all cultures). In this chapter, the advantages and disadvantages of each approach will be discussed.

DOI: 10.4324/9781003051497-13

Adaptation

Adapting a test involves much more than just translating it into the target language (Hambleton, 2005). Because of the cultural bias of some items, it is often necessary to replace or remove items from the new version. After translation, pilot studies should be run in order to check the appropriateness of the test materials. Corrections must be made if original items are regarded as inappropriate and replaced items must be re-tested. Next, it is necessary to perform validity and reliability studies to test if these properties stand across versions and the adapted version is equivalent to the original one. Finally, normative data has to be collected for the target population of the adapted version. It is evident from this description that the adaptation process is long, arduous, and expensive. It demands several steps that might take several years depending on the available resources. The issues with this approach and its associated advantages and disadvantages will be discussed below.

Language

Language is among the most prominent issues in this process. After revision of the test, the next step in the adaptation process is usually the translation. Translation is an art in which the translator must convey the meaning of the original text and not merely find the equivalent words in the target language. However, in neuropsychology, the original words may have been selected not only for their meaning but also for their morphology, i.e., their length, accentuation, or sound. For example, sentence repetition tests are aimed at the measurement of the memory span, thus participants are presented with increasingly longer sentences (Strauss et al., 2006). Word length differs across languages; therefore original words might not be appropriate in the target version of the test. In other cases, the original words might not exist in the target language. For example, words like "shoplifter" or "dive" do not have an exact equivalent in Spanish.

Moreover, it is estimated that 7,117 languages are currently spoken worldwide (Eberhard et al., 2020). As Fernandez & Marcopulos (2019) stated, adapting a basic battery of only five tests into the 23 languages that account for more than half the world's population would involve 115 adaptations. This represents a huge task for even a basic battery of tests, and would result in many languages without any adapted versions (for example, German, Italian, Dutch, Greek). The Repeatable Battery for the Assessment of Neuropsychological Status (RBANS) is probably one of the neuropsychological batteries with the highest number of translations. It has been translated into over 20 languages according to the publishers. Nonetheless, it appears that only a few of these translations (fewer than 10) are adaptations (Fernandez & Marcopulos, 2019).

Some countries are a challenge in terms of developing linguistic/cultural adaptations. India has 22 officially recognized languages, but the inclusion of tribal languages raises this number to more than one thousand (Mallikarjun, 2020). South Africa has 11 official languages. Therefore, there is not an "Indian" or

"South African" version of a certain test. Adapting tests in these countries means producing a version for each linguistic/cultural group.

Copyright

Many neuropsychological tests are copyrighted; therefore those intending to adapt tests need permission from the publishers. Because of the trade-off between the cost of the adaptation and projected sales, publishers are not always interested in adapting a test. Moreover, sometimes permission is granted but the resulting adaptation is not marketed which undermines the utility of this adapted version: researchers who produced the adaptation may not be allowed to distribute it, and thus users are not aware of this version since it is not marketed. Adapting only tests that are not copyrighted is a way to overcome this situation, but copyrighted tests are usually more popular and have been more used in research and clinical settings and thus will often have stronger evidence of their usefulness.

Inappropriateness of items

Some items might be culturally biased which makes them inappropriate for the target language/culture version. For example, the Addenbrooke's Cognitive Examination (ACE) III has a memory item in which the patient is asked to re-member a proper name and address. The original version contains a British name and address. In a review of translated/adapted versions of the ACE III, Mirza et al. (2017) found that, of the 28 papers reporting cultural adaptations, all but one adapted the name/address item. Furthermore, six out of the 19 ACE III sub-tests were frequently adapted and in some versions (e.g., Matias-Guiu et al., 2015) up to ten sub-tests were adapted. This demonstrates that adapting a test demands detailed and arduous work at the level of item revision and replacement.

The challenge of replicating validity studies with pathological groups

Validation of neuropsychological tests often involves a comparison of the per-formance of clinical and control groups in order to provide evidence that the test is sensitive to deficits in the cognitive construct of interest. This kind of study should be run with the target population after a test has been adapted. In some cases the adapted version does not show the same validity evidence as the original version. For example, Fernandez & Fulbright (2015) found that the adapted Argentinian version of the Boston Naming Test had a very low sensitivity which seriously compromised its clinical value.

Replicating validity studies is usually a very demanding task for researchers adapting neuropsychological tests, especially in environments where neu-ropsychology is not well developed. Running a comparison study between clinical and control groups involves having access to clinical populations with a confirmed diagnosis and sometimes the use of brain imaging or other expensive techniques. In some countries, especially those in which adapted tests are needed, neuropsychology is an emerging field and even neurology services are not very common. Moreover, these services are usually focused on delivering clinical services and do not have the resources or training to compile the necessary information to run a scientific study like this. Therefore, access to clinical populations with confirmed diagnosis is very limited and as a consequence running comparison studies between clinical and control samples is not feasible. This frequently leads to a situation in which a test has been adapted but its psychometric properties have not been properly tested. If the adapted version, as in the case of the Argentinian BNT, happens to be a test with unsatisfactory validity evidence, clinicians might be misled by the incorrect information produced by it. What is even more harmful is that clinicians might be unaware that a test is flawed.

The challenge of recruiting large normative samples

Normative data are supposed to be based on samples large enough to control the bias produced by participants with atypical performances. In most cases, neuropsychological tests involve individual administrations. This is a costly, although feasible, procedure for societies with the necessary resources. Nevertheless, for those societies in which research funding is limited and/or scarce, recruiting large normative samples is unlikely or impossible. Kumar & Sadasivan (2016) estimated that there were about 50 practicing neuropsychologists in India, a country with a 1.2 billion population and 22 officially recognized languages. In Botswana there is only one officially recognized neuropsychologist for the whole country (Lingani Mbakile-Mahlanza, personal communication, September 1, 2020). The ratio of neuropsychologists to the national population is another index reflecting the status of neuropsychology as an emerging field in many countries. This ratio has been estimated as 1/540,000 for South Africa, 1/704,000 for Serbia and 1/1,995,250 for Turkey, just to mention a few examples (Grote & Novitski, 2016; Hokkanen et al., 2019).

Despite its continuing growth across the world, it seems that neuropsychology is well developed (i.e., a reasonable ratio of neuropsychologists to national population, several training programs, fair availability of neuropsychological services, adequate number of developed/adapted tests for their local population, sufficient funding for research and sound scientific production in the field) in probably a handful of countries if we consider the 193 states that are members of the United Nations (Ponsford, 2017). In most countries, human and economic resources are insufficient to run expensive normative studies which result in adapted tests without proper normative data.

Development

Developing specific tests for a specific culture as the second option presents almost the same challenges as adapting tests. There are, however, some advantages and disadvantages as compared to adapting existing tests.

Language

This is one of the major advantages of developing tests as compared to adapting. Because the test is planned from the beginning to be used in a specific language, the complicated process of translation, finding linguistic equivalents, and testing the new items can be avoided. All the time dedicated to translation can be saved with this approach.

Copyright

This is another advantage of this option. There is no need to ask for permission to adapt the test which reduces time and cost. In addition, the authors have the freedom to commercially or freely distribute the test.

Marketing/distribution

The downside of not working with a copyrighted test is the challenges of marketing and distribution. Some tests developed for some specific cultures in the appropriate language are unknown by many local neuropsychologists, and as a consequence, they are not frequently used. This undermines the very essence of developing a test: to be used, either by clinicians or researchers. All the effort of developing a test might be fruitless because of its poor dissemination. Related to this, one of the great challenges is balancing the cost of producing a test (if specific materials are required), cost of advertising in order that people know about the test and cost of distribution with the desire to ensure that tests are as widely and cheaply available as possible.

Inappropriateness of items

As in the case of language, cultural inappropriateness of items can be avoided since items are specifically developed for a specific culture. This, again, represents a saving in time and effort.

Loss of accumulated experience

This is one of the great disadvantages of developing over adapting. Many neuropsychological tests have existed for decades and have been used in numerous research studies. There are multiple demonstrations of their validity and reliability. This is very valuable information as it conveys confidence in the data they yield. The adaptation process, if successful, takes advantage of that accumulated experience with the test; however, all this experience is lost with the option of developing a new test to measure the same construct.

The challenge of demonstrating the psychometric properties of the test and recruiting large normative samples

Developing a new test does not represent any advantage over adapting an existing one as regards testing its psychometric properties (validity, reliability) and collecting normative data. Both approaches involve testing these psychometric properties and are equally demanding in each case. The challenges of gathering normative data in environments with very limited resources have been discussed above. An alternative solution, although modest from a methodological perspective, is to develop cut-off scores (Fernandez & Abe, 2018). With this approach in validity studies comparing control and clinical samples, a cut-off score might be derived as a reference for clinicians. This approach cannot take advantage of the subtleties of using normative data, i.e., to determine with higher precision the quality of the performance of a specific participant in comparison with his/her reference group. It is limited more to focusing only on identifying the level of performance on the test that indicates significant impairment. Another major limitation of this approach is that it does not typically take account of demographic-related performance variance (e.g., performance that is affected by age, gender, education etcetera). Nonetheless, it can provide the clinicians with useful information establishing the odds of the participant of being classified as positive or negative as regards a certain condition.

Time extension

Developing a new test involves designing new items and, in some cases, new procedures. These processes might take a long time because new items need to be tested before developing the final version of the test. New items may not be appropriate for many reasons (e.g., being difficult to understand for participants, lack of specificity, not belonging to the intended cognitive domain, etcetera). Depending on the available resources it might take months or even years to develop the stimuli for a new test.

Although adapting tests may also require the development of new items (which may not work as well as the original items) it is likely that one may have greater confidence in items selected to perform in a similar way to established items and so require less time for development studies.

Development of cross-cultural tests

CCTs are designed from the outset to be used in different cultural settings. There is experience spanning around 40 years in the development of this sort of test (Fernandez & Marcopulos, 2019). However, in the last decade efforts in this area have started to produce better results. As an alternative to adapting existing tests or developing new culture-specific tests, it is important to analyze the pros and cons of this approach as well.

Ease of the translation/adaptation process

The major advantage of this approach is that this kind of test is designed specifically to be appropriate for many cultures. These tests use stimuli that are recognized by participants from a wide variety of cultures. For example, the Cross-Cultural Dementia Screening (CCD) uses the sun and the moon as stimuli for its Stroop test version (Goudsmit et al., 2017). These stimuli can be recognized by participants of any culture in the world. Although some translation/adaptation may still be required, the fact that these tests are designed to be suitable for many cultures, makes them easily adaptable to different cultures. Most likely they will only need a translation of verbal stimuli and instructions. For example, the RUDAS has been rapidly translated into around a dozen languages in a very short span of years (Araujo et al. 2018; Fernandez & Marcopulos, 2019; Komalasari et al., 2019; Nepal et al., 2019). Most of these versions were simply translated or needed very minor adaptations.

The small number of tests developed using this model

Although efforts toward designing CCTs started in the 1980s there have been relatively few tests developed under this model and many of them are not in use currently (Fernandez & Marcopulos, 2019). In addition, most are dementia screening tests. As a result, most of these tests are not appropriate for young populations or children and do not go beyond a superficial exploration of cognitive functions. The CCTs currently in use do not provide an in-depth assessment of some areas of cognition such as executive functions. Although some efforts to address these populations and cognitive functions have started to emerge – such as the European Cross-Cultural Neuropsychological Test Battery (CNTB) (Nielsen et al., 2018a, b), or the Multicultural Neuropsychological Scale (Fernandez et al., 2018) there is still a lot of work to do to satisfy the needs in this field.

Minor adaptations might be necessary

Despite the explicit intention to use rather universal stimuli, the items of these tests might not be completely free of cultural bias. For example, the RUDAS has been translated to most languages without changes in its stimuli; however, the Nepali version needed an adaptation since 85% of their sample was unable to draw the cube. They substituted the cube drawing for a stick design task (Nepal et al., 2019). Thus, the aspiration to design tests based on items and tasks that will work in *all* cultures may not be fully achievable, but nevertheless, the extent of adaptation required may be substantially less for CCT's than for our existing tests.

Need to develop local norms and test psychometric properties

Even when the use of stimuli common to many cultures represents a huge advantage over adapting existing tests there is still a need to develop local norms and test the psychometric properties of the translated version. Unfortunately, because of the difficulties entailed in the development of norms and other psychometric studies, many translations of CCT's just present the adaptation without testing the reliability, validity or even providing local norms. For example, in their RUDAS review, Komalasari et al. (2019) found that only three out of 12 studies of RUDAS adaptation reported validity and reliability data. As previously discussed, this is not enough to ensure that the test is effective in the new context.

The CNTB is one of the rare exceptions providing normative data that represent various cultural groups: Belgian, Danish, German, Greek, Norwegian, Swedish, Moroccan, Pakistani/Indian Punjabi, Polish, Turkish, and Yugoslavian. They also provide validity data (Nielsen et al., 2018a,b). This remarkable accomplishment involved the participation of several European research centers, mostly from Western Europe, and was funded by the European Union. This reinforces the view that these enterprises demand the availability of a significant number of resources (human and economic).

An alternative approach used in some cases is the development of cut-off scores. As mentioned before in this chapter, this is a methodologically limited solution but preferable to the complete lack of psychometric information.

Conclusion

Each approach described here, adapting existing tests, developing local tests or CCTs, has advantages and disadvantages. There is not a simple answer to what is the preferable option. It seems that each one of these options should be considered carefully in every particular situation. The best option might be different in every case. For example, if a language test evaluating many aspects (i.e., grammar, phonology, comprehension, etc) is necessary, then developing a local test for the target language seems to be the best option. However, if an executive functioning test widely used is easy to adapt to the target culture considering that the stimuli are appropriate to it, then adapting seems the best option. Finally, developing CCTs might help cover the assessment of basic cognitive functioning in several cultures at the same time.

Unfortunately, many times decisions about these options are based on circumstantial factors such as the particular interest of the researchers, familiarity with a specific test, or unawareness of other test options. A careful consideration of goals, resources, and feasibility should be performed in order to address the need for neuropsychological tests in contexts where existing tests are not available.

References

Araujo, N.B., Nielsen, T.R., Engedal, K., Barca, M.L., Coutinho, E.S., & Laks, J. (2018). Diagnosing dementia in lower educated older persons: Validation of a Brazilian Portuguese version of the Rowland Universal Dementia Assessment Scale (RUDAS). *Brazilian Journal of Psychiatry, 40(3)*, 264–269. 10.1590/1516-4446-2017-2284.

Ardila, A. (2007). The impact of culture on neuropsychological test performance. In B.P. Uzzell, M. Pontón, & A. Ardila (Eds.), *International handbook of cross-cultural neuropsychology* (pp. 23–44). Mahwah: Lawrence Erlbaum Associates.

Eberhard, D.M., Simons G.F., & Fennig C.D. (Eds.). (2020). *Ethnologue: Languages of the World. Twenty-third edition.* Dallas, Texas: SIL International. Online version: http://www.ethnologue.com

Fernandez, A.L. & Abe, J. (2018). Bias in cross-cultural neuropsychological testing. Problems and possible solutions. *Culture & Brain, 6,* 1–35.

Fernandez, A.L. & Fulbright, R.L. (2015) Construct and Concurrent Validity of the Spanish Adaptation of the Boston Naming Test. *Applied Neuropsychology: Adult, 22*(5), 355–362. DOI: 10.1080/23279095.2014.939178.

Fernandez, A.L., Jáuregui Arriondo, G., Folmer, M., Seita, V., Ciarímboli, G. & Aimar, C. (2018). Development of the multicultural neuropsychological scale (MUNS): A new tool for neuropsychological assessment of culturally diverse populations. *The International Annals of Medicine, 2*(8). 10.24087/iam.2018.2.8.594.

Fernandez, A.L., & Marcopulos, B.A. (2019). Cross-cultural tests in neuropsychology: A review of recent studies and a modest proposal. In S. Koffler, E.M. Mahone, B. Marcopulos, D. Johnson-Greene, & G. Smith. *Neuropsychology: A Review of Science and Practice III* (pp. 93–128). New York: Oxford University Press Series.

Goudsmit, M., Uysal-Bozkir, Ö., Parlevliet, J.L., van Campen, J.P., de Rooij, S.E., & Schmand, B. (2017). The cross-cultural dementia screening (CCD): A new neuropsychological screening instrument for dementia in elderly immigrants. *Journal of Clinical and Experimental Neuropsychology, 39*(2), 163–172. 10.1080/13803395.2016.12 09464.

Grote, C.L., & Novitski, J.I. (2016) International perspectives on education, training, and practice in clinical neuropsychology: comparison across 14 countries around the world. *The Clinical Neuropsychologist, 30*(8), 1380–1388. DOI: 10.1080/13854046.2016.1235 727.

Hambleton, R. (2005) Issues, designs, and technical guidelines for adapting tests into multiple languages and cultures. In R.K. Hambleton, P.F. Merenda & C.D. Spielberger (Eds.), *Adapting educational and psychological tests for cross-cultural assessment* (pp. 3–38). Mahwah: Lawrence Erlbaum Associates.

Hokkanen, L., Lettner, S., Barbosa, F., Constantinou, M., Harper, L., Kasten, E., Mondini, S., Persson, B., Varako, N., & Hessen, E. (2019). Training models and status of clinical neuropsychologists in Europe: Results of a survey on 30 countries. *The Clinical Neuropsychologist, 33*(1), 32–56. 10.1080/13854046.2018.1484169.

Komalasari, R., Chang, H.C.R., & Traynor, V. (2019). A review of the Rowland Universal Dementia Assessment Scale. *Dementia (London, England), 18*(7–8), 3143–3158. 10.1177/1471301218820228.

Kumar, J.K., & Sadasivan, A. (2016). Neuropsychology in India. *The Clinical Neuropsychologist, 30*(8), 1252–1266. 10.1080/13854046.2016.1197314.

Mallikarjun, B. (2020). Linguistic demography of the tribal languages in India. *Language in India, 20*(6), 120–140.

Matias-Guiu., J.A., Fernández de Bobadill, R., Escudero, G., Pérez-Pérez, J., Cortés, A., Morenas-Rodríguez, E., Valles-Salgado, M., Moreno-Ramos, T., Kulisevsky, J., & Matías-Guiu, J. (2015). Validation of the Spanish version of Addenbrooke's cognitive examination III for diagnosing dementia. *Neurologia, 30*(9),545–551.

Mirza, N., Panagioti, M., Waheed, M.W., & Waheed, W. (2017). Reporting of the translation and cultural adaptation procedures of the Addenbrooke's Cognitive

Examination version III (ACE-III) and its predecessors: A systematic review. *BMC Medical Research Methodology*, *17*(*1*), 1–10. 10.1186/s12874-017-0413-6.

Nell, V. (2000). *Cross-cultural neuropsychological assessment: Theory and practice*. Mahwah, NJ: Lawrence Erlbaum Associates.

Nepal, G.M., Shrestha, A., & Acharya, R. (2019). Translation and cross-cultural adaptation of the Nepali version of the Rowland universal dementia assessment scale (RUDAS). *Journal of Patient-Reported Outcomes*, *1*. 10.1186/s41687-019-0132-3.

Nielsen, T.R., Segers, K., Vanderaspoilden, V., Beinhoff, U., Minthon, L., Pissiota, A., … Waldemar, G. (2018a). Validation of a European cross-cultural neuropsychological test battery (CNTB) for evaluation of dementia. *International Journal of Geriatric Psychiatry*, *34*(*1*), 144–152. 10.1002/gps.5002.

Nielsen, T.R., Segers, K., Vanderaspoilden, V., Bekkhus-Wetterberg, P., Minthon, L., Pissiota, A., … Waldemar, G. (2018b). Performance of middle-aged and elderly European minority and majority populations on a cross-cultural neuropsychological test battery (CNTB). *The Clinical Neuropsychologist*, *32*(*8*), 1411–1430. 10.1080/13854046.2 018.1430256.

Ponsford, J. (2017). International growth of neuropsychology. *Neuropsychology*, *31*(*8*), 921–933. 10.1037/neu0000415.

Strauss, E., Sherman, El, & Spreen, O. (2006). *A compendium of neuropsychological tests: Administration, norms and commentary* (3rd. Ed.). New York: Oxford University Press.

United Nations, Department of Economic and Social Affairs. Population Division. (2019). *International migrant stock 2019* (United Nations database. POP/DB/MIG/Stock/ Rev.2019).

11

INTERPRETER-ASSISTED NEUROPSYCHOLOGICAL ASSESSMENT: CLINICAL CONSIDERATIONS

Daryl Fujii, Octavio Santos, and Lori Della Malva

Introduction

International migration has been steadily climbing, with an estimated 272 million migrants living abroad in 2019. Roughly half of them reside in nine countries: the United States (51 million), Germany and Saudi Arabia (13 million each), Russia (12 million), United Kingdom (10 million), United Arab Emirates (nine million), France, Canada and Australia (around eight million each), and Italy (six million). The regions with the highest percentage of migrants within the total population are Oceania (21.2%) and North America (16.0%). The primary reasons for migration are labor, family reunification, and asylum. The leading countries of origin for migrants are India (18 million), Mexico (12 million), China (11 million), Russia (10 million), and the Syrian Arab Republic (eight million) (United Nations, 2019).

A significant challenge for migrant countries is providing access to healthcare, including psychological services. As many migrants do not speak or are not fluent in the language of the host country, a major barrier to healthcare is the lack of interpreters and appropriate language services (Satinsky et al., 2019). Interpreters are similarly needed to provide access to healthcare services in multilingual countries, such as India where there are 23 official languages and 415 living languages (New World Encyclopedia, 2020). In general, an interpreter may be required when: (a) the patient is not proficient in the host language; (b) language demands of the meeting are high; and (c) fluency may be affected by acute illness, injury or medical crisis, dementia, or emotionally charged situations (APS, 2013; CISOC, 2013). The use of interpreters is crucial for competent healthcare services, as it is associated with lower risk for medical errors; better healthcare utilization clinical outcomes and patient satisfaction; reduced miscommunication; facilitating self-disclosure; and increased diagnostic accuracy in mental health services (Bauer

DOI: 10.4324/9781003051497-14

& Alegría, 2010; Gonçalves et al., 2013; Hudelson et al., 2012; Jacobs et al., 2010; Mehler et al., 2004).

The importance of using interpreters in psychological services is supported by ethical guidelines of some national psychological associations. Such guidelines indicate that services must be provided in a language the patient understands to ensure informed consent and appropriate treatment, and the psychologists' responsibility for using trained interpreters or training paraprofessionals if interpreter services are not available (APA, 2017; CPA, 2017). If training ad hoc interpreters, clinicians should select persons who are fluent in the language of the clinician and patient, and possess an understanding of the two different cultural contexts (BPS, 2017). Guidelines generally argue against using children or other family members as interpreters, as this practice can result in problems with confidentiality, place family members in uncomfortable roles and may undermine their relationships, and frequently results in inaccurate translations (APA, 2017; APS, 2013; BPS, 2017; CISOC, 2013; Tribe & Thompson, 2011).

Despite the strong evidence for the clinical utility of interpreters, countries differ widely in their availability of interpreter services and their utilization when available (Kluge et al., 2012; Ledoux et al., 2018). Instead, many clinicians will opt for ad hoc interpreters, including family members due to cost and scheduling issues (AHA, 2018; Bischoff & Hudelson, 2010; Gray & Hardt, 2017; Hadziabdic & Hjelm, 2019; Satinsky et al., 2019). Factors impacting the availability of interpreters include laws or policies supporting equal access for healthcare, government funding or insurance coverage, chronology of accepting migrants, and availability of speakers in a given language (Bischoff & Hudelson, 2010; Kluge et al., 2012). Another issue is variability in interpreter training standards across countries. Although countries such as Sweden have general community interpreter training (Gustafsson et al., 2012), specialized interpreter certification programs for healthcare are only offered in a handful of countries, such as Australia, Canada, the U.K., and the U.S. (Souza, 2020). In other countries, such as Denmark, formal interpreter training is available but is not required (Skammeritz et al., 2019). A related issue is familiarity with the specific content areas. For example, a medically trained interpreter may not be familiar with the process, issues, and terminology for clinical neuropsychology.

The purpose of this chapter is to provide guidance for using interpreters during neuropsychological assessments for patients who do not speak the neuropsychologist's language. The goal is to facilitate communication between the neuropsychologist and patient to maximize fairness in conducting the evaluation. First, we will describe cultural standards for fairness in testing and preparations for ensuring fairness, with an emphasis on language considerations. Then, guidelines for working with interpreters, including pre-assessment preparation, assessment procedures, and post-assessment debriefing will be described. Finally, clinical considerations for using interpreters will be illustrated with considerations for persons with hearing impairment.

Cultural standards and preparations for fairness in testing

Psychological testing is based on western behavioral assumptions and values. Thus, it may be biased for people from cultures with differing experiences and values (Greenfield, 1997). To address this, the American Education Research Association identified four standards that must be met to maximize test fairness for culturally diverse test-takers: (a) test-takers must be comfortable with the testing situation; (b) tests must have construct and content validity for the test-takers; (c) test-takers must be able to comprehend test items and respond appropriately; and (d) tests must be valid for their intended purpose (AERA, 2014).

To maximize fairness in testing, neuropsychologists must have a good knowledge base of the patient's culture, especially if different than their own, as it provides a context for understanding behaviors. This knowledge base should then guide strategies for the assessment, data interpretation and conceptualization, and recommendations. A model for understanding the impact of culture on neuropsychological assessment is the ECLECTIC Framework, whose acronym stands for pertinent cultural facets that can aid in conceptualizing the patient: Education and literacy, Culture, Language, Economics, Communication style, Testing situation, Intelligence conception, and Context of immigration (for a review, see Fujii, 2018). The information from these cultural facets can be applied to formulate a testing strategy, which would include working with an interpreter. For example, cultural knowledge would be crucial for developing rapport, conceptualizing how the patient may view the testing situation, educating the patient to alleviate concerns, and maximizing communication. The latter would entail an appreciation of communication style which are language pragmatics, such as when to talk, what can be disclosed and to whom, and idioms of distress or alternative modes of expressing or manifesting distress related to personal and cultural meaning (for a review, see Tannen, 1984).

Mismatches in communication styles between persons cannot only result in miscommunication but also impede rapport (Tannen, 1984). For example, in cultures with a direct communication style, information is imparted in "what is said" while the assumed responsibility for communication is with the speaker. Thus, for individuals in western countries, it is important to be explicit when requesting to ensure the listener understands what is needed. By contrast, for cultures with indirect communication styles, a large part of the message is in "what is *not* said", and the responsibility for communication is assumed to be with the listener. For example, if a Japanese person responds to a question by nodding with a silent smile, the questioner is expected to understand that either the answer is "no" or the Japanese person is not comfortable with the question. Thus, knowledge of communication style differences between the neuropsychologist and patient is essential, as it provides additional context for understanding what the patient verbalizes. Another important role for an interpreter is being a cultural broker for the clinician, thus identifying cultural or social factors that might impact the encounter (Kasten et al., 2020; Tribe & Thompson, 2011).

Cultural knowledge would also be important for selecting appropriate tests, scrutinizing tests for item biases, determining language(s) spoken, and the need for an interpreter for a particular language. For instance, when preparing for a Filipino patient, cultural knowledge and a pre-assessment interview would help the neuropsychologist determine if the evaluation should be conducted in Tagalog, Ilocano, Visayan, English, or a combination of languages. This determination would be important for test selection or the need for translating test instructions beforehand; the latter should only be a last resort and not be attempted for conceptual verbal tests (Casas et al., 2012; Roger & Code, 2011) unless translations adhere to existing guidelines (International Test Commission, 2017). Any translation issues during the evaluation should be reported. Finally, cultural knowledge also guides the selection and appropriateness of norms and validity of test findings, which would be essential for appropriate data interpretation and treatment recommendations. Cultural facets, such as the existence of a written language, quality of education or access to the Internet or books, would significantly impact opportunities for learning, cognitive development, and neuropsychological test performance (for a review, see Fujii, 2018).

Conducting an interpreter-assisted evaluation

Selecting in-person or remote interpreters

Interpreting is a skilled role and, whenever possible, psychologists are strongly encouraged to use accredited interpreters when conducting assessments (APS, 2013). Professionally trained interpreters adhere to an ethics code as well as maintain high professional conduct standards and competence, thus reducing the risk of compromising confidentiality or information being withheld or distorted in translation because of family relationships or the emotional/sensitive nature of the issue (Zhang & Wang, 2019). Access to remote interpreter services, including video and phone interpreters, can ensure assistance when in-person interpreters are immediately unavailable. Currently, there is only a paucity of studies comparing the effectiveness of remote versus in-patient interpreter services. In general, patients report greater satisfaction with in-person and video interpreters versus telephone interpreter services (Joseph et al., 2017; Schulz et al., 2015); the latter may be associated with clinically relevant language omissions, additions, and substitutions, suggesting potential for hindering care and relationship building (Lor & Chewning, 2016). In-person interpreter services have been rated higher than video interpreter services for self-assessed cultural competence (Nápoles et al., 2010).

General pre-assessment considerations

The following are a suggested checklist and recommendations for clinical neuropsychologists based on a review of guidelines and guides on the use of interpreters for assessment purposes (APA, 2017; APS, 2013; BPS, 2017; CISOC, 2013; Tribe & Thompson, 2011):

- Familiarize yourself with culturally appropriate methods and materials to conduct your assessment in addition to securing interpretation services.
- Determine language and/or dialect spoken in the patient's country of origin. Some languages are spoken in many countries (e.g., Arabic) and dialectal variations can significantly affect communication.
- Consider that interpreter-mediated assessments typically take longer and adjust your schedule accordingly.
- Consider the layout of the room, testing requirements, and seating arrangements. For assessment purposes, seating the interpreter at an equidistant position between the clinician and the patient, with the clinician facing the patient to allow eye-to-eye contact is recommended.
- The interpreter's presence creates a triad, so be mindful of the potential shift in dynamic and how it will be addressed.
- Consider creating written guidelines and/or a contract for the interpreter.
- Meet with the interpreter before the evaluation to establish rapport and a good working relationship, as the latter is also beneficial for the patient. In that meeting:
  - Take reasonable steps to ensure that no prior non-professional relationship exists between the patient and the interpreter.
  - Ask the interpreter to avoid being alone with the patient before and after the session and reinforce that non-session-related conversations with the client should be avoided.
  - Familiarize the interpreter with the nature and objectives of the session.
  - Review your respective roles, boundaries, and relevant ethical points, stressing that it is the clinician who controls the proceedings.
  - Stress the importance of accuracy and the need for as exact as possible interpretation of the client's responses.
  - Discuss pertinent cultural issues.
  - Clarify special terminology and complex concepts that may be discussed.
  - Discuss practical aspects, such as seating arrangement, mode of interpretation (e.g., consecutive), the pace of delivery of information, and signals to be used if clarification or pauses are required by either you or the interpreter.
  - If applicable, forewarn the interpreter that clients may have difficulty expressing themselves clearly and coherently, or that the information conveyed may make little sense.
  - If a discussion of distressing events may be anticipated, brief the interpreter regarding the possibility that content may be distressing. Interpreters may be better prepared to manage the traumatic nature of a meeting if they are advised that they might find it upsetting.

Preparing for administration of psychometric measures

- Caution is recommended in the use and interpretation of psychometric tests when working with an interpreter, as an informal translation of an instrument created in one language may alter the meaning and level of difficulty of the items, and possibly result in inaccurate scores. Psychometric assessment tools should be discussed with the interpreter during the pre-session briefing. To improve the validity of scores, consider the following:

  - Discuss the appropriateness of measures within the language and cultural context.
  - Agree on signals the interpreter can feel at ease to use if they have difficulty translating a word or phrase
  - Discuss the need to translate psychometric tests in advance. Tests with high verbal conceptual load should be avoided. Translations of instructions for tests requiring simple exemplars are more appropriate. The translation should be checked for parity of meaning and cognitive load with the original item rather than requesting spontaneous translation during consultation.
  - Explain standardized assessment procedures to the interpreter and allocate some time to practice test administration.
  - Ask the interpreter not to give any additional assistance to service users during psychometric testing.
  - Request verbatim translation of responses.

Conducting the assessment

Consent and limits of confidentiality

- Patients should be advised ahead of the assessment that an interpreter will be present.
- The interpreter should arrive before the client and leave after the client has left.
- Introduce yourself and the interpreter.
- Speak directly to the patient, maintaining eye contact when speaking.
- Patients may sometimes refuse an interpreter. Common concerns typically revolve around personal trust and confidentiality, or issues surrounding an interpreter's gender, religion, or ethnic background. Patients may also believe their language proficiency is sufficient and an interpreter is not required. If you believe otherwise, the following may help address the concerns of the patient:

  - Inform the patient that the interpreter is a trained professional whose sole role is to facilitate communication.
  - Inform the patient that the interpreter is bound to a code of ethics that includes confidentiality and impartiality.

- Determine whether giving the patient a choice of interpreter gender and nationality (if feasible) could alleviate concerns.
- If the patient feels their language proficiency is adequate, you may suggest using the interpreter as a backup only if communication problems arise.
- If the client insists no interpreter is needed and you believe otherwise, explain that you need to have an interpreter to ensure your own understanding.

- Explain your respective roles and the purpose of the meeting, clearly stating that you alone hold clinical responsibility for the meeting and describe the boundaries of confidentiality.
- Explain that you and the interpreter are both bound by ethical and professional standards to maintain confidentiality.
- Make efforts to ensure a comfortable atmosphere where all involved feels able to ask for clarification
- Explain that the interpreter cannot participate in the conversation beyond the interpretation and that the patient has questions or does not understand something, s/he should tell you not the interpreter.
- Reiterate that everything will remain confidential and linguistically interpreted, even in side conversations.
- Obtain informed consent from the patient to use the interpreter and verify that the client/patient can understand the interpreter and is ready to proceed.

During the evaluation

- Ask the interpreter to interpret everything said by the patient.
- Conversation between you and the interpreter that does not require translation should be avoided.
- Speak in a normal tone, using straightforward and simple language.
- Be aware of pace; pause between every 2-3 sentences and between thoughts, and do not break up thoughts.
- Remind the patient that s/he should pause frequently to let the interpreter translate.
- Use repetition, examples and visual aids, and write out numbers or multiple-step instructions.
- Avoid jargon, humor, complex grammar, and compound questions.
- Be mindful of signals from the interpreter while keeping your focus on the patient.
- Allow time for the interpreter to process and convey information. Be aware of times where cultural interpretation is required (i.e., interpreter uses knowledge of the client's culture to elucidate a point or clarify a misunderstanding).
- Be mindful that it can become easy to lose concentration or to lose the thread of the session with the slower pace of interpreter-mediated assessments.

- Periodically check-in with the patient to ensure you have understood each other well.
- Ask the client to rephrase anything you do not understand.
- At the end of the session, summarize and review the meeting and consider inviting the patient to provide feedback about the experience.

Post-assessment de-brief

- After the assessment has been completed, you may find a short de-brief meeting with the interpreter informative.
- Consider soliciting feedback from the interpreter about what worked and what may be improved in the future.
- Discuss whether there were any difficulties with the translation of items, make a note of these, and consider the implications on the validity of the results for those items.
- Share perceptions and observations and clarify any additional cultural issues.
- Consider the interpreter's psychological wellbeing and provide a short, informal debriefing discussion following consultation with a client in the event of potential distress or vicarious traumatization

Considerations for test interpretation and report writing

- There is much heterogeneity within the population of every country. Thus, neuropsychologists must interpret data within the context of the individual's social status, education level, and developmental history. For example, a trilingual child of a diplomat would be expected to perform better on cognitive tests than a monolingual child of a rural farmer who is illiterate.
- Make direct reference to the issue of neuropsychological and cognitive assessment difficulties when working with non-host language-speaking service users and using interpreters.
- Consider the impact of psychometric limitations and culture on conclusions and formulation. Psychometric limitations include, but are not limited to, administering the test(s) in a patient's second or third language, interpreter-mediated assessments, and lack of translated, culturally adapted, validated, and normed tests for the individual's population. Cultural considerations include, but are not limited to, economic resources associated with access to a quality education, the internet, and reading material; experiences and environmental demands impacting cognitive styles, concept of intelligent behavior, and common medical and psychiatric illnesses; and sociopolitical history impacting world view, values, and social and behavioral norms, including the sense of self, expectations for self-disclosure, communication style, and idioms of distress (Fujii, 2018). In these cases, neuropsychologists should not base clinical decisions strictly on test scores and the patient's presentation interpreted within the context of western culture. Instead, neuropsychologists must gauge the

potential impact of these factors on test data when determining formulation and diagnoses and describe one's confidence in the conclusions.

Neuropsychological assessment of persons with Hearing Impairment or Deafness (HID)

There are 466 million people worldwide with HID (World Health Organization, 2018). In many clinical settings in which neuropsychologists practice, rates of HID are significantly higher than in the general population (Garahan et al., 1992). Persons with HID require specialized assessment (AERA, 2014). When testing persons with HID, testing standards and nondiscrimination policies for persons with disabilities require neuropsychologists to ensure fairness and accuracy in test administration and interpretation (AERA, 2014; Hill-Briggs et al., 2007). Appropriate test selection must be based on both demographics and an understanding of task demands. Demographic factors related to deafness that should be considered include but are not limited to (Braden, 1994; Hill-Briggs et al., 2007; Tekin et al., 2001): (a) whether HID onset occurs before 18 months of age or after language development; (b) etiology and presence of neurological or physical co-morbidities; (c) progressive versus non-progressive hearing loss; and (d) parental hearing status.

Neuropsychological assessment with persons with HID presents significant challenges, including but not limited to selecting tests, modifying test administration to accommodate disability, and interpreting results from nonstandard test administration (Hill-Briggs et al., 2007). For a signed paired associates test and other commonly used tests with HID and their limitations, the reader is referred to Hill-Briggs et al. (2007) and Pollard et al. (2005). When a neuropsychologist fluent in the person's preferred language/communication mode is unavailable to conduct the assessment, a certified interpreter with experience in mental health interpreting should be used (Vernon & Miller, 2001), even if the person generally talks well in an informal setting. However, simply interpreting a test into sign language does not necessarily make it accessible to deaf persons or test the same ability that it does with hearing persons. For example, written instructions are inappropriate accommodation as reading skills are highly variable and generally low in this population. If the person responds orally, caution must be used to avoid scoring articulation errors. Most deaf persons do have some residual hearing, so auditory distractions should be avoided. Similarly, visual distractions are more problematic for a deaf person, so they should be minimized. Caution must be used when selecting, administering, and interpreting personality tests that appear to be "language free" (i.e., reading not required) and, if interpretation is used, make sure that all information, such as non-manual cues (e.g., facial expression), is included (Hill-Briggs et al., 2007).

While an interpreter must be able to sign or use Cued Speech (augmented lip-reading with the help of hand-coding) accessible to the patient, the neuropsychologist should understand the nature of interpreting (Vernon & Miller,

2001). For example, if the task is to interpret from the English stimuli to American sign language (ASL), the interpreter must alter the stimuli. Given that there are words without sign equivalents, the interpreter must either fingerspell the English word or explain the word, thus conveying the interpreter's understanding of the word. Since the interpreter often must cognitively process sentences or instructions before conveying them to the deaf person, this may place some task demands on the interpreter rather than the person. The interpreter also must adjust information to make it accessible in the context of the person's culture in which case more information may be needed to ensure understanding. Direct interpretation of test stimuli also can often alter tasks for some measures. For instance, administering a sentence memory task via ASL signs with the same word order in English to retain the task structure increases task difficulty for a person with ASL as a primary language. Similarly, if the sentences are interpreted to ASL, their length and complexity are likely lost, making their scoring also difficult. The interpretation of test questions sometimes provides too much information (e.g., body parts are indicated by the speaker pointing to their own body part to indicate the patient to do the same). Modeling or extra practice trials may be needed to ensure task understanding. On perceptual-motor tasks, asking a deaf person to close their eyes or wear a blindfold is problematic, as it cuts them off from communication. Allowing the patient to provide written responses may result in an apparent English "word salad" given ASL word order (Hill-Briggs et al., 2007).

In sum, neuropsychologists working with deaf persons should be aware of demographic, communication, cultural and differential neurological/neurodevelopmental factors that impact neuropsychological assessment administration, performance, and interpretation of results. Research is necessary for the development of appropriate testing methods and insurance of reliable and valid neuropsychological assessment practices for individuals with HID.

Summary

An implication of steady increases in international migration is the need for interpreter services to provide competent healthcare to migrants, which includes the neuropsychological assessment. Studies clearly support the clinical advantage of trained interpreters and neuropsychologists should strive to utilize them when performing an assessment. However, the availability of certified interpreters and training standards vary across countries, and ad hoc interpreters are commonly used to provide health services. This chapter attempted to address the current practice reality by providing a cultural context for assessing patients who speak a different language from the clinician and detailed recommendations for working with and training interpreters for a neuropsychological assessment. Implementation was illustrated with specific considerations for working with persons with HID.

References

AERA. (2014). *Standards for educational and psychological testing.* American Educational Research Association, American Psychological Association, and National Council on Measurement in Education.

AHA. (2018). *AHA Hospital Statistics, 2018 Edition.* American Hospital Association. https://www.aha.org/statistics/2016-12-27-aha-hospital-statistics-2018-edition.

APA. (2017). *Ethical principles of psychologists and code of conduct.* American Psychological Association. https://www.apa.org/ethics/code/ethics-code-2017.pdf.

APS. (2013). *Working with interpreters: A practice guide for psychologists.* Australian Psychological Society Professional Practice. http://docplayer.net/6406225-Working-with-interpreters-a-practice-guide-for-psychologists-aps-professional-practice.html.

Bauer, A.M., & Alegría, M. (2010). Impact of patient language proficiency and interpreter service use on the quality of psychiatric care: A systematic review. *Psychiatric Services,* *61*(8), 765–773. 10.1176/ps.2010.61.8.765.

Bischoff, A., & Hudelson, P. (2010). Access to healthcare interpreter services: Where are we and where do we need to go? *International Journal of Environmental Research and Public Health,* *7*(7), 2838–2844.

BPS. (2017). *Working with interpreters: Guidelines for psychologists.* British Psychological Society. https://www.bps.org.uk/news-and-policy/working-interpreters-guidelines-psychologists.

Braden, J.P. (1994). *Deafness, deprivation, and IQ.* Plenum Press.

Casas, R., Guzmán-Vélez, E., Cardona-Rodriguez, J., Rodriguez, N., Quiñones, G., Izaguirre, B., & Tranel, D. (2012). Interpreter-mediated neuropsychological testing of monolingual spanish speakers. *Clinical Neuropsychologist,* *26*(1), 88–101. 10.1080/13854 046.2011.640641.

CISOC. (2013). *Guide to working with interpreters.* Cultural Interpretation Services for Our Communities. http://www.cisoc.net/_files/Guide_to_Working_with_Interpreters.pdf.

CPA. (2017). *Canadian code of ethics for psychologists - fourth edition.* Canadian Psychological Association. 10.1037/h0086812.

Fujii, D.E.M. (2018). Developing a cultural context for conducting a neuropsychological evaluation with a culturally diverse client: the ECLECTIC framework*. *Clinical Neuropsychologist,* *32*(8), 1356–1392. 10.1080/13854046.2018.1435826.

Garahan, M.B., Waller, J.A., Houghton, M., Tisdale, W.A., & Runge, C.F. (1992). Hearing loss prevalence and management in nursing home residents. *Journal of the American Geriatrics Society,* *40*(2), 130–134. 10.1111/j.1532-5415.1992.tb01932.x.

Gonçalves, M., Cook, B., Mulvaney-Day, N., Alegría, M., & Kinrys, G. (2013). Retention in mental health care of Portuguese-speaking patients. *Transcultural Psychiatry,* *50*(1), 92–107. 10.1177/1363461512474622.

Gray, B., & Hardt, E.J. (2017). A comparison of the use of interpreters in New Zealand and the US. *New Zealand Medical Journal,* *130*(1456), 70–75.

Greenfield, P. M. (1997). You can't take it with you: Why ability assessments don't cross cultures. *American Psychologist,* *52*(10), 1115–1124. 10.1037/0003-066x.52.10.1115.

Gustafsson, K., Norström, E., & Fioretos, I. (2012). Community interpreter training in spoken languages in Sweden. *International Journal of Interpreter Education,* *4*(2), 24–38.

Hadziabdic, E., & Hjelm, K. (2019). Register-based study concerning the problematic situation of using interpreting service in a region in Sweden. *BMC Health Services Research,* *19*(1). 10.1186/s12913-019-4619-7.

Hill-Briggs, F., Dial, J.G., Morere, D.A., & Joyce, A. (2007). Neuropsychological assessment of persons with physical disability, visual impairment or blindness, and hearing impairment or deafness. *Archives of Clinical Neuropsychology*, *22*(3), 389–404. 10.1016/j.acn.2007.01.013.

Hudelson, P., Perneger, T., Kolly, V., & Perron, N. (2012). Self-assessed competency at working with a medical interpreter is not associated with knowledge of good practice. *PLoS ONE*, *7*(6), e38973. 10.1371/journal.pone.0038973.

International Test Commission. (2017, April 3). *The ITC guidelines for translating and adapting tests*. International Journal of Testing; Routledge. 10.1080/15305058.2017.1398166.

Jacobs, E.A., Diamond, L.C., & Stevak, L. (2010). The importance of teaching clinicians when and how to work with interpreters. *Patient Education and Counseling*, *78*(2), 149–153. 10.1016/j.pec.2009.12.001.

Joseph, C., Garruba, M., & Melder, A. (2017). Patient satisfaction of telephone or video interpreter services compared with in-person services: Asystematic review. *Australian Health Review*, *42*(2), 168–177.

Kasten, M.J., Berman, A.C., Ebright, A.B., Mitchell, J.D., & Quirindongo-Cedeno, O. (2020). Interpreters in health care: A concise review for clinicians. *American Journal of Medicine*, *133*(4), 424–428.e2. 10.1016/j.amjmed.2019.12.008.

Kluge, U., Bogic, M., Devillé, W., Greacen, T., Dauvrin, M., Dias, S., Gaddini, A., Koitzsch Jensen, N., Ioannidi-Kapolou, E., Mertaniemi, R., Puipcinós i Riera, R., Sandhu, S., Sarvary, A., Soares, J.J.F., Stankunas, M., Straßmayr, C., Welbel, M., Heinz, A., & Priebe, S. (2012). Health services and the treatment of immigrants: Data on service use, interpreting services and immigrant staff members in services across Europe. *European Psychiatry*, *27*(SUPPL.2). 10.1016/S0924-9338(12)75709-7.

Ledoux, C., Pilot, E., Diaz, E., & Krafft, T. (2018). Migrants' access to healthcare services within the European Union: A content analysis of policy documents in Ireland, Portugal and Spain. *Globalization and Health*, *14*(1). 10.1186/s12992-018-0373-6.

Lor, M., & Chewning, B. (2015). Telephone interpreter discrepancies: Vvideotapes of Hmong medication consultations. *International Journal of Pharmacy Practice*, *24*(1), 30–39.

Mehler, P.S., Lundgren, R.A., Pines, I., & Doll, K. (2004). A community study of language concordance in Russian patients with diabetes. *Ethnicity and Disease*, *14*(4), 584–588.

Nápoles, A. M., Santoyo-Olsson, J., Karliner, L. S. O'Brien, H., Gregorich, S.E., & Pérez-Stable E.J. (2010). Clinician ratings of interpreter mediated visits in underserved primary care settings with ad hoc, in-person professional, and video conferencing modes. *Journal of Health Care for the Poor and Underserved*, *21*(1), 301–317.

New World Encyclopedia. (2020). *Languages of India*. https://www.newworldencyclopedia.org/entry/Languages_of_India.

Pollard, R.Q., Rediess, S., & DeMatteo, A. (2005). Development and validation of the Signed Paired Associates Test. *Rehabilitation Psychology*, *50*(3), 258–265. 10.1037/0090-5550.50.3.258.

Roger, P., & Code, C. (2011). Lost in translation? Issues of content validity in interpreter-mediated aphasia assessments. *International Journal of Speech-Language Pathology*, *13*(1), 61–73. 10.3109/17549507.2011.549241.

Satinsky, E., Fuhr, D.C., Woodward, A., Sondorp, E., & Roberts, B. (2019). Mental health care utilisation and access among refugees and asylum seekers in Europe: A systematic review. *Health Policy*, *123*(9), 851–863. 10.1016/j.healthpol.2019.02.007.

Schulz, T. R., Leder, K., Akinci, I., & Biggs, B. A. (2015). Improvements in patient care: Videoconferencing to improve access to interpreters during clinical consultations for refugee and immigrant patients. *Australian Health Review*, *39*(4), 395–399.

Skammeritz, S., Sari, N., Jiménez-Solomon, O., & Carlsson, J. (2019). Interpreters in transcultural psychiatry. *Psychiatric Services*, *70*(3), 250–253. 10.1176/appi.ps.201800107.

Souza, I.E.T.d.V. (2020). The Medical Interpreter Mediation Role: Through the Lens of Therapeutic Communication. In I.E.T.d.V. Souza & E. Fragkou (Eds.), *Handbook of Research on Medical Interpreting* (pp. 99–135). IGI Global.

Tannen, D. (1984). The pragmatics of cross-cultural communication. *Applied Linguistics*, *5*(3), 189–195. 10.1093/applin/5.3.189.

Tekin, M., Arnos, K.S., & Pandya, A. (2001). Advances in hereditary deafness. *Lancet*, *358*(9287), 1082–1090. 10.1016/S0140-6736(01)06186-4.

Tribe, R., & Thompson, K. (2011). Developing guidelines on working with interpreters in mental health: Opening up an international dialogue? *International Journal of Culture and Mental Health*, *4*(2), 81–90. 10.1080/17542863.2010.503365.

United Nations. (2019). *The number of international migrants reaches 272 million, continuing an upward trend in all world regions, says UN.* https://www.un.org/development/desa/en/news/population/international-migrant-stock-2019.html.

Vernon, M., & Miller, K. (2001). Interpreting in mental health settings: Issues and concerns. *American Annals of the Deaf*, *146*(5), 429–434. 10.1353/aad.2012.0200.

World Health Organization. (2018). *Deafness.* https://www.who.int/news-room/facts-in-pictures/detail/deafness.

Zhang, J., & Wang, L. (2019). Education on qualified interpreters – strategies of cultivation on psychological quality. *Sino-US English Teaching*, *16*(7), 310–314. 10.17265/1539-8072/2019.07.005.

12

THE INFLUENCE OF ACCULTURATION ON NEUROPSYCHOLOGICAL TEST PERFORMANCE

Yi Wen Tan and Gerald H. Burgess

Introduction

Many neuropsychological tests are developed in the West, and subsequently, cultural theorists assert that tests are biased toward Westernized interpretations of intellect and cognitive abilities (Ardila, 2007; Helms, 1992; Nell, 2000). Consequently, it is difficult to gauge the extent to which tests measure ability or idiosyncratic thinking styles in different cultures (Helms, 1992). A culturalist's position would assert that immersion into a different culture than one's own can influence performance on neuropsychological tests (Ardila, 2007; Helms, 1992; Nell, 2000), and since tests are purported to reflect Western values, culturalists posit that greater immersion or *acculturation* of non-Western individuals into a Western culture ought to lead to better test performance (Helms, 1992; Nell, 2000; Van de Vijver et al., 1999).

What is acculturation?

The process of acculturation

Acculturation is described as a process of change due to continual contact between individuals of different cultures (Redfield et al., 1935). Graves (1967) distinguished between group-level acculturation (i.e., social norms and socio-political structures) and individual or psychological acculturation (i.e., affect, behavior, cognitions). This distinction is found in Berry's (1997, 2017) model of acculturation, where the process occurs in stages (Figure 12.1). First, psycho-sociocultural characteristics at the collective level, like inter-racial attitudes, migration policies, and economic conditions in both the dominant and non-dominant society initiate the process of acculturation. Pre-existing individual-level factors like personality traits, motivation,

DOI: 10.4324/9781003051497-15

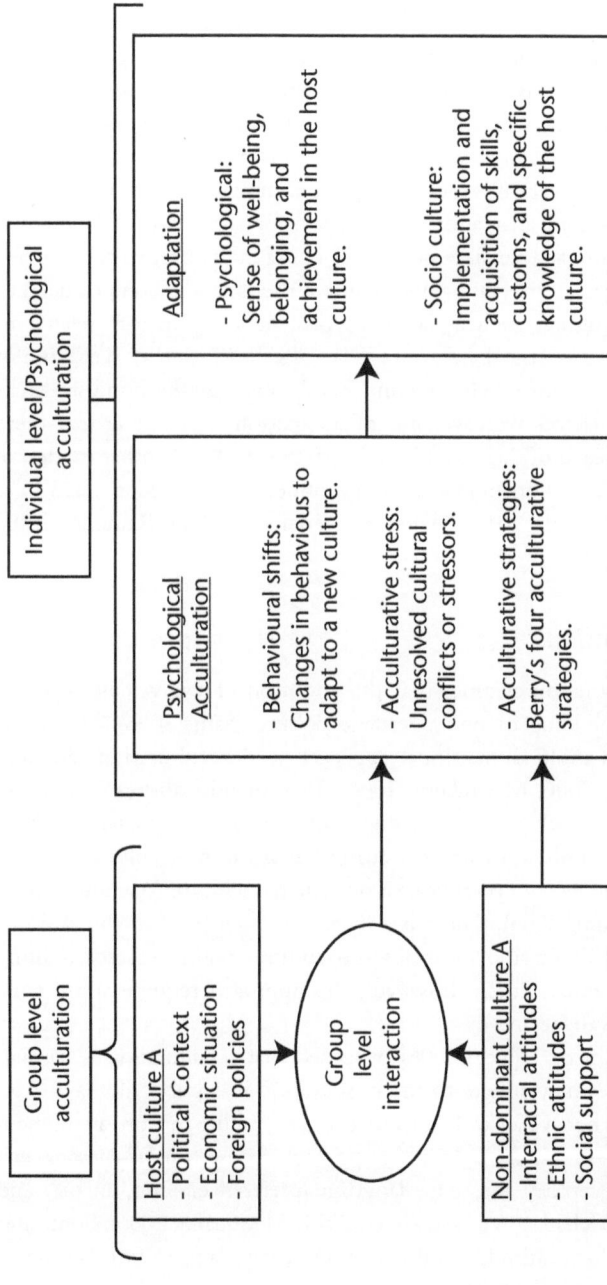

FIGURE 12.1 Berry's model of acculturation (adapted from Berry, 2017).

or self-efficacy also govern the process (Berry, 1997). Over time, behavioral and psychological shifts at the individual level serve as a function for adapting to the host's society. Acculturative stress, typically associated with psychiatric disorders, may emerge if the individual cannot resolve negative experiences with a new culture (i.e., discrimination, prejudice, etc.; Berry, 1997; Williams & Berry, 1991). At the second stage, Berry (1997) proposed four acculturative strategies, which explain how an individual orients themself in relation to the new culture. These include (a) assimilation, where an individual adopts the host culture while shedding their native culture; (b) separation, which is the rejection of the host culture to maintain one's own culture; (c) integration, where both cultures are equally maintained, or bi-culturalism; and (d) marginalization, where the individual rejects both the host and their heritage cultures. Adaptation is the last stage of acculturation with two streams (Berry, 2017). Psychological adaptation is characterized as a sense of belonging, personal achievement, and an overall sense of well-being in the host culture. Socio-cultural adaptation is characterized as having acquired the skills and competencies to function in the host culture, such as language proficiency, primarily related to managing daily functions in the host society. Berry's (1997, 2017) work is not without limitations; for instance, Schwartz and Zamboanga (2008) found six latent factors, two additional factors were variants of integration, concluding that there were more variants for each of these acculturative strategies. While other models of acculturation exist, Berry's work remains highly influential and widely cited over various fields (Ferguson et al., 2012; Peeters & Oerlemans, 2009; Rudmin, 2003; Yoon et al., 2013).

Domains and dimensions

Unidimensional acculturation assumes that the adoption of a new culture is synonymous with the shedding of one's heritage culture (Sam, 2006). However, there is a shift toward multi-dimensional approaches in acculturation (Arends-Tóth & van de Vijver, 2006; Matsudaira, 2006). Bi-dimensionality is commonly found in the literature, with two independent dimensions: (a) *adoption*, which refers to how much an individual absorbs attributes of the host culture, and (b) *maintenance*, refers to the level of retention of one's native culture (Arends-Tóth & van de Vijver, 2006; Sam, 2006). There is a recent emergence of tri-dimensional models, where a third-dimension measures sub-cultures within society (Chung et al., 2004; Ferguson et al., 2012), however, this approach requires more conceptual and empirical validation.

Arends-Tóth and van de Vijver (2006) proposed that acculturation is *domain-specific*. Ethnic minorities may adhere to the host culture in a public domain (e.g., work, school, etc.), but behave according to their native culture in a private domain (e.g., home, religious settings, etc.). Another approach focuses on *life-domains,* encompassing facets of everyday life, e.g., language preference, food, media, and friendship preferences (Celenk & Van de Vijver, 2011; Matsudaira, 2006). Some also distinguish between values, attitudes, and behaviors (Matsudaira, 2006). Behaviors are overt, like participation in cultural activities or language skills (Arentoft et al., 2012). Attitudes are thoughts and feelings held toward the host or one's native culture, such as Berry's acculturative strategies (Arends-Tóth & van de Vijver, 2006). Cultural values are subjective internal belief systems, such as ethnic identity or cultural attachment (Matsudaira, 2006). Moreover, these domains are also dimensional, for example, the domain of language is measured for the host language (adoption dimension) and native language (maintenance dimension). Importantly, these domains are conceptually independent. Proficiency in the host culture's language will not automatically produce strong social affiliation to the host culture (Arends-Tóth & van de Vijver, 2006; Matsudaira, 2006; Schwartz et al., 2010). Therefore, a single index of acculturation – be it a composite score or reliance on a single domain like language, cannot account for the complexity of acculturation (Arends-Tóth & van de Vijver, 2006; Matsudaira, 2006; Schwartz et al., 2010).

Acculturation and neuropsychological test performance

Our research has revealed that the impact of acculturation on test performance is not clear in the neuropsychological literature (Tan et al., 2020). Furthermore, there does not appear to be consistency in the way neuropsychological tests and acculturation are defined in the literature. Some authors discuss neuropsychological assessments as a conglomerate of all standardized instruments including scholastic tests (Helms, 1992; Nell, 2000). Others rely on a single construct like language proficiency (Harris et al., 2013), or proxy measures of acculturation, such as years of residency (Boone et al., 2007), as surrogates of acculturation. This is despite the theoretical literature advocating that a single proxy estimate cannot account for the complexity of acculturation (Arends-Tóth & van de Vijver, 2006; Lopez-Class et al., 2011). Furthermore, there is very little empirical work about its clinical relevance. We found that the association between scales of acculturation and test performance is inconsistent across different tests, empirical studies, and samples (Tan et al., 2020). This was still the case with stringent criteria that adjusted for confounding effects of age and education. We found that there was no discernible pattern as to *what* type of tests would be related to constructs of acculturation, with the exception of performance on tests of visual and verbal delayed memory. These tests consistently did not associate with acculturative measures (Tan et al., 2020). Other than this, acculturation did not associate with all types of tests that are theoretically labeled under the same cognitive construct. For instance, the "language" domain of an acculturative measure predicted Weschler's VCI' Vocabulary subtest but not the Similarities subtest (Hasson et al., 2019; Razani et al., 2007). From the literature overall we concluded that the relationship between test performances and acculturation seemed to depend on the specific study's methodology (e.g., type of measures, method of analyses, power and effect size), and sample characteristics (e.g., representativeness of sample, clinical versus non-clinical samples) within each study. Although the relationship between test performance and acculturation appeared weak, some plausible trends also emerged in our investigations, and these will be highlighted over the next few sections. In any case, it is important to note that various component constructs within acculturation (i.e., domains, dimensions, stages, etc.) can exhibit different predictive values for different types of tests, which we will now turn to.

Measuring acculturation for neuropsychology

Arends-Tóth and van de Vijver (2006) outlined several guidelines for scale selection. This section will place some of these in a neuropsychological context where appropriate. It is important to consider the rationale for using scales of acculturation (Arends-Tóth & van de Vijver, 2006). Researchers and practitioners must account for the significance of acculturation for their target population, along with acculturation's practical and theoretical relevance. For research, a clear research question should guide scale selection and the use of specific constructs within acculturation. We do not go as far as recommending an acculturation scale, but we strongly recommend scrutinizing the psychometric properties of scales, and only use scales that have undergone stringent validity tests (e.g., construct, criterion validity, etc.).

Domains and dimensions for test performance

Most studies of acculturation rely on outdated unidimensional scales (Tan et al., 2020), but some evidence suggests that higher "adoption" is generally related to better test performance. Evidence for the effects of "maintenance" on test performances however is comparatively weaker. Arentoft et al. (2012) explored the relationship between a range of cognitive indices with acculturation in a sample of HIV+ Latin Caribbeans. They found that "maintenance" predicted lower scores on an index of executive function comprised of the Wisconsin Card Sorting Test (WCST) and the Trail Making Test-B (TMT-B) (Arentoft., et al., 2012). Within the same study, the strength of association was weaker for maintenance compared to adoption (Arentoft et al., 2012). Tan and Burgess (2018) examined the influence of four domains of adoption (i.e., language competency, knowledge, food preferences, cultural identity) on the Short Parallel Assessments of Neuropsychological Status (SPANS; Burgess, 2014). The SPANS consists of seven internally themed, multiple item cognitive indices (i.e., orientation, attention, language, visuospatial skills, memory and learning, mental efficiency, and cognitive flexibility) (Burgess, 2014). Only the domains of language and cultural knowledge (i.e., knowledge of customs, traditions, etc.) adoption, predicted tests of language and orientation-to-political-leadership respectively for a healthy ethnically diverse group in the UK. This finding coincided with the wider literature, suggesting that language adoption was more likely to exert a stronger influence on test performance compared to other domains of acculturation (Tan et al., 2020). For example, Hasson et al. (2019) demonstrated that among three domains of adoption and maintenance (language use, ethnic identity, cultural knowledge), only language adoption is associated with the WAIS-IV Vocabulary subtest and WAIS-VIQ.

Adjusting for language variables (i.e., the language of administration, language status) may "partial-out" the effects of acculturation on tests (Tan et al., 2020). Most studies however have relied on a single index of acculturation (Tan et al., 2020). In the current state of literature, there is insufficient data on the effects of multiple dimensions, particularly the less linguistically driven domains of acculturation (i.e., ethnic identity, attitudes, etc.) on test performances. Nonetheless, our research pointed to the possibility that language and culturally specific knowledge, within adoption, are likely important to test performances in similarly labeled neuropsychological test domains.

Models of acculturation and neuropsychology

It is important to consider how models of acculturation integrate with neuropsychological research. We will explain the nature of the relationship between performance and acculturation using Berry's model of acculturation. At the first stage, individual differences, like self-efficacy or personality factors, could be precursors for the relationship between later stages of acculturation and test performances (Berry, 1997). Next, accounting for group-level acculturation can broaden research beyond test performances. For instance, it might be interesting to consider how constructs at this stage affect the access, relevance, and national health policies related to neuropsychological practice. This can include service evaluations involving racial attitudes, immigration policies, or inter-ethnic relations on streams of healthcare like neurorehabilitation.

At the second stage, individual or psychological acculturation includes acculturative strategies and acculturative stress. Marginalized ethnic minorities, with high levels of acculturative stress, typically experience worse health outcomes and poorer cognitive functioning (Kennepohl et al., 2004; Williams & Berry, 1991). Our review however failed to find detailed studies for the relationship between acculturative strategies and test performances (Tan et al., 2020). Nguyen et al. (2012) found that higher levels of acculturative stress were related to lower performance on the TMT and the Benton Visual Retention Test. However, Tan (2020) found no relationship between acculturative stress and a battery of neuropsychological tests that purport to measure similar constructs. Generally, empirical work regarding the relationship between psychological acculturation and test performance is scarce in the literature. Rather, our research suggested that the final stage of Berry's model is likely to be most salient for neuropsychological performance (Tan, 2020; Tan et al., 2020). We found that language and cultural knowledge are more likely to possess greater predictive value for test performances, and these coincide with socio-cultural adaptation in Berry's model.

We acknowledge that there are other models of acculturation but advocating for one model over the other is beyond our scope. Nonetheless, integrating models of acculturation with neuropsychological theories allows stronger bonds between empirical and theoretical evidence, and extends the scope of neuropsychological research. Such an approach is rarely found in the literature (Arends-Tóth & van de Vijver, 2006; Tan, et al., 2020). If a multi-stage model of acculturation is used, we recommend exploring how neuropsychological variables differ at various stages of acculturation.

Types of scales

Most scales in the literature measure acculturative behaviors, e.g., language skills, cultural participations, etc. (Celenk & Van de Vijver, 2011; Matsudaira, 2006). Only a handful measure attitudes, and even fewer assessed values (Matsudaira, 2006). Language skills or preference is the most common life-domain, followed by food, media, and friendship preferences (Celenk & Van de Vijver, 2011; Matsudaira, 2006). In our investigation, a large proportion of scales used in neuropsychological studies were biased toward the language domain (Tan et al., 2020). For example, ethnic identity was measured according to language preferences, neglecting the uniqueness of different domains of acculturation. While scales should cover multiple domains, there must be a conceptual and empirical distinctions for each domain.

Scales developed for specific ethnic groups are not evenly distributed. There exist several scales for Hispanics, African Americans, Asian Americans, but few are available for other ethnic groups outside the US (Celenk & Van de Vijver, 2011; Matsudaira, 2006). Some scales are purported to fit multiple ethnic groups, such as the Abbreviated Multidimensional Acculturation Scale (Zea et al., 2003). However, opinions are mixed over the generalizability of the same scale across different ethnicities. Generic scales may not capture subtle acculturative nuances between and within ethnic groups (Rudmin, 2009). Others assert that underlying factors of acculturation are transferrable across cultures (Arends-Toth & Van de Vivjer, 2006). Another issue would be whether immigrant and non-immigrant populations can be measured equally. Later generations (i.e., non-immigrant groups) potentially face unique cultural challenges. They may reject their native culture, resulting in intergenerational conflicts within their community, leading to acculturative stress (Miller et al., 2011; Williams & Berry, 1991). However, some studies demonstrate that acculturative scales are invariant across generations (Huynh et al., 2018; Miller & Lee, 2009; Miller et al., 2011). Just because a scale is valid for one specific group does not necessarily mean it is generalizable to all populations of that ethnic group. We strongly recommend calculating psychometric properties of scales, even if it received previous validation to ensure consistency of results.

Proxies of acculturation

It is incorrect to assume that proxy variables (i.e., years of residency, generational status, etc.) are synonymous with acculturation (Arends-Tóth & van de Vijver,

2006; Matsudaira, 2006). These proxies should be measured separately with different research aims from acculturation. In fact, studies have found that these proxies have different associative values on test performance from acculturation. For instance, years of residency predicted WAIS-IV Similarities in Arab Americans (Hasson et al., 2019), generational status predicted the California Verbal Learning Test and TMT-A in Japanese Americans (Kemmotsu et al., 2013), but these tests did not associate with measures of acculturation. As such, it is important to use direct measures of acculturation as opposed to relying on proxy variables.

Clinical implications

Few studies have explored the clinical implications of acculturation on test performances. In relation to theory, we found that *adaptation* in Berry's model seems more salient toward test performances. Subsequently, clinicians might account for language abilities and cultural knowledge of the "host" culture during routine examinations. However, the extent and the way these constructs affect clinical decisions and interpretations of test results are unclear. In the literature recommendations for the use of acculturation scales are mixed (Tan et al., 2020). Some studies advocate for the use of scales during clinical examination, but these lack details over how such scales could be used, nor provide sufficient evidence to back up purported recommendations. Currently, simply accounting for the construct of acculturation is insufficient to 'correct' for confounding factors that arise during testing. Other cultural (i.e., language of administration, ethnicity, etc.) and non-cultural factors (i.e., motivation, test-wiseness, etc.) may confound, and to a seemingly greater extent, test performances independent of acculturation (Tan, 2020).

Acculturation and test norms

Existing guidelines lack clarity over how scales ought to be used in clinical situations. However, utilizing ethnic and demographically-corrected norms may attenuate the effects of acculturation on tests (Tan, 2020). For instance, groups with high and low unidimensional acculturation scores were significantly different on raw scores on the WCST, but this finding disappeared when Hispanic norms were used (Krch et al., 2015). Arentoft et al. (2012) found significant correlations between acculturation and several tests, but Mindt et al. (2014) could not replicate these with a different set of norms. Demographically correct norms could reduce cultural biases, consequently, diminish the effects of acculturation on performances. More clinically relevant, Simpao et al. (2005) found that lower language proficiency and lower social affiliation toward English-speaking American culture predicted a higher likelihood of scoring below a clinical cut-off on the Mini-Mental Status Exam. These investigations suggest that scales might be used to assess whether norms or clinical thresholds from a cultural majority are transferrable to minority groups.

Tan (2020) attempted to directly explore the predictive value of acculturation with British English-speaking norms on the SPANS (Burgess, 2014). Language adoption predicted the likelihood of attaining a false positive (>1 SD below the mean) on an index of language for an ethnically-diverse healthy sample, and cultural knowledge for a Political Leadership subtest of orientation. However, other cultural measures could perform equally well in predicting false positives in this study. Tan (2020) used a cultural quotient (CQ; Burgess, 2014), assessing for English as first language status, proportion of lifetime residing in an English-speaking country, and whether English was taught during formative years of education. The CQ predicted the likelihood of attaining false positives for almost the same set of tests as acculturation did. While it is possible to use acculturative scales to determine the generalizability of existing norms for diverse populations, other measures like historic experience with the host language appear just as useful. As such, the current literature does not provide enough detailed information regarding the practicality or the justification for using acculturative measures in clinical practice.

Summary of recommendations

Tan et al. (2020) provide an in-depth discussion regarding the relationship between acculturation and test performances. We further recommend works by Arends-Tóth and van de Vijver (2006) as they outlined important considerations, like item selection, and how domains or dimensions should be measured. We also recommend two systematic reviews to aid scale selection (Celenk & Van de Vijver, 2011; Matsudaira, 2006). Below summarizes recommendations for measuring acculturation in a neuropsychological context.

1. **Rationale for measuring acculturation**
   There must be a clear purpose for measuring acculturation, assessing its relevance for a target population, and its practical and theoretical implications. On a clinical front, neurological impairment can hinder comprehension and some patients are prone to fatigue. Shorter scales with straightforward items could be used (e.g., Bidimensional Acculturation Scale; Marin & Gamba, 2011), but this runs the risk of undermining the complexity of acculturation. Practitioners must balance practical concerns with empirical and theoretical evidence and justify the use of acculturative scales – be it to assess the suitability of existing norms or other clinical purposes.

2. **Contextualize models of acculturation**
   Few studies considered the importance of using an acculturative framework to understand neuropsychological issues. This would strengthen empirical findings with theory and extend the scope of neuropsychological research. For this, constructs of acculturation should be mapped onto theoretical frameworks of acculturation. For example, we used the Asian American Multidimensional Acculturation Scale (Chung et al., 2004) in our study (Tan & Burgess, 2018).

We opined that the language and cultural knowledge subscale was most representative of socio-cultural adaptation in Berry's framework of acculturation.

3. **Scale selection and acculturative constructs**

The framework of acculturation and research aims should guide scale selection and specific constructs. Our data suggest that within socio-cultural adaptation, language and cultural knowledge adoption ought to be more salient to test performances. However, there is a myriad of constructs that could be explored. We recommend moving away from unidimensional scales to using a wide range of domains, as different acculturative constructs can have different predictive values. Next, it is vital to scrutinize the distinctiveness of subscales and report psychometric properties even if it was previously validated. If appropriate, researchers can assess whether a scale captures the unique cultural characteristics of a target population. Lastly, we strongly discourage researchers from assuming that proxy variables are interchangeable with acculturation.

4. **Relative importance of acculturation**

Evidently, there could be other factors important toward test performances compared to acculturation. Cultural and non-cultural variables could have a larger impact on test performances compared to acculturation, or perhaps mediate the relationship. Future researchers could go beyond adjusting for age and education and consider a wider array of variables that could also affect test performance.

Conclusion

We have presented evidence about the current state of the literature on the relationship between acculturation and neuropsychological performance. This relationship however has not yet received sufficient attention, and thus most of our conclusions are tentative. We conclude that within socio-cultural adaptation, language and knowledge adoption are more likely to influence test performance. However, it is still uncertain what type of neuropsychological tests would be more susceptible to which constructs of acculturation. There is also insufficient data to inform how acculturation information should or could be used in clinical practice. Using acculturative scales to determine the transferability of existing norms for diverse populations may be possible, but more work is required to clarify this. Therefore, we are not able to provide any solid clinical recommendations. However, researchers can widen the clinical scope of acculturation, for example, how acculturation affects referrals, neurorehabilitation outcomes, and therapeutic alliance is largely missing in the literature. Acculturation is a diverse construct, but neuropsychological theories are equally multifaceted. In any case, it is important for researchers and clinicians to identify specific constructs in each field to provide clarity over how these two may integrate or interact.

References

Ardila, A. (2007). The impact of culture on neuropsychological test performance. In B.P. Uzzell, M. Ponton, & A. Ardila (Eds.), *International handbook of cross-cultural neuropsychology* (pp. 23–44). Mahwah, NJ: Lawrence Erlbaum Associates.

Arends-Tóth, J., & van de Vijver, F. (2006). Assessment of psychological acculturation. In D.L. Sam & J.W. Berry (Eds.), *The Cambridge handbook of acculturation psychology* (pp. 142–160). New York: Cambridge University Press.

Arentoft, A., Byrd, D., Robbins, R.N., Monzones, J., Miranda, C., Rosario, A., … Rivera Mindt, M. (2012). Multidimensional effects of acculturation on English-language neuropsychological test performance among HIV+ Caribbean Latinas/os. *Journal of Clinical and Experimental Neuropsychology, 34*(8), 814–825.

Berry, J.W. (1997). Immigration, acculturation, and adaptation. *Applied Psychology, 46*(1), 5–34.

Berry, J.W. (2017). Theories and Models of acculturation. In S.J. Schwartz, & J. Unger (Eds.), *The Oxford handbook of acculturation and health* (p. 15). Oxford: Oxford University Press.

Boone, K.B., Victor, T.L., Wen, J., Razani, J., & Pontón, M. (2007). The association between neuropsychological scores and ethnicity, language, and acculturation variables in a large patient population. *Archives of Clinical Neuropsychology, 22*(3), 355–365.

Burgess, G. (2014). *Short Parallel Assessments of Neuropsychological Status (SPANS) manual.* Oxford: Hogrefe.

Celenk, O., & Van de Vijver, F.J. (2011). Assessment of acculturation: Issues and overview of measures. *Online Readings in Psychology and Culture, 8*(1), 10.

Chung, R.H., Kim, B.S.K., & Abreu, J.M. (2004). Asian American multidimensional acculturation scale: Development, factor analyses, reliability, and validity. *Cultural Diversity and Ethnic Minority Psychology, 10*(1), 66–80.

Ferguson, G.M., Bornstein, M.H., & Pottinger, A.M. (2012). Tridimensional acculturation and adaptation among Jamaican adolescent–mother dyads in the United States. *Child Development, 83*(5), 1486–1493.

Graves, T.D. (1967). Psychological acculturation in a tri-ethnic community. *South-Western Journal of Anthropology, 23*, 337–350.

Harris, G., Tulsky, S., & Schultheis, T. (2013). Assessment of the non-native English speaker: Assimilating history and research findings to guide clinical practice. In D.S. Tulsky, D.H. Saklofske, R.K. Heaton, R. Bornstein, M.F. Ledbetter, G.J. Chelune, … A. Prifitera (Eds.), *Clinical interpretation of the WAIS-III and WMS-III* (pp. 343–390). Elsevier Ltd.

Hasson, R., Wu, I., & Fine, J. (2019). Clinical utility of the WASI-II and its association with acculturation levels among Arab American adolescent males. *Applied Neuropsychology: Child*, 1–12.

Helms, J.E. (1992). Why is there no study of cultural equivalence in standardized cognitive ability testing? *American Psychologist, 47*(9), 1083–1101.

Huynh, Q.L., Benet-Martínez, V., & Nguyen, A.M.D. (2018). Measuring variations in bicultural identity across US ethnic and generational groups: Development and validation of the Bicultural Identity Integration Scale – Version 2 (BIIS-2). *Psychological Assessment, 30*(12), 1581.

Kemmotsu, N., Enobi, Y., & Murphy, C. (2013). Performance of older Japanese American adults on selected cognitive instruments. *Journal of International Neuropsychological Society, 19*(7), 773.

Kennepohl, S., Shore, D., Nabors, N., & Hanks, R. (2004). African American acculturation and neuropsychological test performance following traumatic brain injury. *Journal of the International Neuropsychological Society, 10*(4), 566–577.

Krch, D., Lequerica, A., Arango-Lasprilla, J.C., Rogers, H.L., DeLuca, J., & Chiaravalloti, N.D. (2015). The multidimensional influence of acculturation on Digit Symbol-Coding and Wisconsin Card Sorting Test in Hispanics. *The Clinical Neuropsychologist, 29*(5), 624–638.

Lopez-Class, M., Castro, F.G., & Ramirez, A.G. (2011). Conceptions of acculturation: A review and statement of critical issues. *Social Science & Medicine, 72*(9), 1555–1562.

Matsudaira, T. (2006). Measures of psychological acculturation: A review. *Transcultural Psychiatry, 43*(3), 462–487.

Marin, G., & Gamba, R.J. (1996). A new measurement of acculturation for Hispanics: The Bidimensional Acculturation Scale for Hispanics (BAS). *Hispanic Journal of Behavioral Sciences, 18*(3), 297–316.

Miller, M.J., Kim, J., & Benet-Martínez, V. (2011). Validating the riverside acculturation stress inventory with Asian Americans. *Psychological Assessment, 23*(2), 300.

Miller, M.J., & Lee, R.M. (2009). Factorial invariance of the Asian American Family Conflicts Scale across ethnicity, generational status, sex, and nationality. *Measurement and Evaluation in Counseling and Development, 42*(3), 179–196.

Mindt, M.R., Miranda, C., Arentoft, A., Byrd, D., Monzones, J., Fuentes, A., … Morgello, S. (2014). Aging and HIV/AIDS: Neurocognitive implications for older HIV-positive Latina/o adults. *Behavioral Medicine, 40*(3), 116–123.

Nell, V. (2000). *Cross-cultural neuropsychological assessment: Theory and practice.* Mahwah, NJ: Lawrence Erlbaum Associates.

Nguyen, H.T., Quandt, S.A., Grzywacz, J.G., Chen, H., Galván, L., Kitner-Triolo, M.H., & Arcury, T.A. (2012). Stress and cognitive function in Latino farmworkers. *American Journal of Industrial Medicine, 55*(8), 707–713.

Peeters, M.C., & Oerlemans, W.G. (2009). The relationship between acculturation orientations and work-related well-being: Differences between ethnic minority and majority employees. *International Journal of Stress Management, 16*(1), 1.

Redfield, R., Linton, R., & Herskovits, M.J. (1935). A memorandum for the study of acculturation. *Man, 35*, 145–148.

Razani, J., Murcia, G., Tabares, J., & Wong, J. (2007). The effects of culture on WASI test performance in ethnically diverse individuals. *The Clinical Neuropsychologist, 21*(5), 776–788.

Rudmin, F. (2009). Constructs, measurements and models of acculturation and acculturative stress. *International Journal of Intercultural Relations, 33*(2), 106–123.

Rudmin, F.W. (2011). Acculturation, acculturative change, and assimilation: A research bibliography with URL links. *Online Readings in Psychology and Culture,* Unit 8.

Rudmin, F. W. (2003). Critical history of the acculturation psychology of assimilation, separation, integration, and marginalization. *Review of General Psychology, 7*(1), 3–37.

Sam, L. (2006). Acculturation: Conceptual background and core components. In L. Sam & J.W. Berry (Eds.), *The Cambridge handbook of acculturation.*

Schwartz, S.J., Unger, J.B., Zamboanga, B.L., & Szapocznik, J. (2010). Rethinking the concept of acculturation. *American Psychologist, 65*(4), 237–251.

Schwartz, S.J., & Zamboanga, B.L. (2008). Testing Berry's model of acculturation: A confirmatory latent class approach. *Cultural Diversity and Ethnic Minority Psychology, 14*(4), 275.

Simpao, M.P., Espino, D.V., Palmer, R.F., Lichtenstein, M.J., & Hazuda, H.P. (2005). Association between acculturation and structural assimilation and Mini-Mental State Examination–Assessed cognitive impairment in Older Mexican Americans: Findings from the San Antonio Longitudinal Study of Aging. *Journal of the American Geriatrics Society, 53*(7), 1234–1239.

Tan, Y.W. (2020). *The relationship between acculturation and neuropsychological test performances* (Unpublished doctoral dissertation). Leicester, United Kingdom: University of Leicester.

Tan, Y.W., & Burgess, G.H. (2018). Multidimensional effects of acculturation at the construct or index level of seven broad neuropsychological skills. *Culture and Brain*, 1–19.

Tan, Y.W., Burgess, G.H., & Green, R.J. (2020). The effects of acculturation on neuropsychological test performance: A systematic literature review. *The Clinical Neuropsychologist*.

Tsai, J.L., Ying, Y.W., & Lee, P.A. (2000). The meaning of "being Chinese" and "being American" variation among Chinese American young adults. *Journal of Cross-Cultural Psychology, 31*(3), 302–332.

Van de Vijver, F.J., Helms-Lorenz, M., & Feltzer, M.J. (1999). Acculturation and cognitive performance of migrant children in the Netherlands. *International Journal of Psychology, 34*(3), 149–162.

Williams, C.L., & Berry, J.W. (1991). Primary prevention of acculturative stress among refugees: Application of psychological theory and practice. *American Psychologist, 46*(6), 632.

Yoon, E., Chang, C.T., Kim, S., Clawson, A., Cleary, S.E., Hansen, M., ... & Gomes, A.M. (2013). A meta-analysis of acculturation/enculturation and mental health. *Journal of Counselling Psychology, 60*(1), 15.

Zea, M.C., Asner-Self, K.K., Birman, D., & Buki, L.P. (2003). The Abbreviated Multidimensional Acculturation Scale: Empirical validation with two 208 Latino/Latina samples. *Cultural Diversity and Ethnic Minority Psychology, 9*(2), 107.

13

NEUROPSYCHOLOGICAL REHABILITATION: PERSPECTIVES BASED ON CULTURAL EXPERIENCE

Jill Winegardner

Rehabilitation has been defined as a *two-way interactive process* whereby survivors of brain injury work together with professional staff and others to achieve their optimum physical, psychological, social, and vocational well-being (McLellan 1991). This definition is not universally accepted, though, and my experience has shown that in neuropsychological rehabilitation, as in all rehabilitation, the culture of the person with disability and of their environment is of prime importance and must play an integral role in both assessment and interventions. In this chapter, I will first review the literature on this subject and then describe my own experience of the impact of culture on neuropsychological rehabilitation while working as a neuropsychologist in four countries. This experience demonstrates how the culture of a country in which the person with disability lives – its wealth, infrastructure, and social attitudes to disability – has a profound effect on if, how and with what aims neuropsychological rehabilitation can take place.

Review of published literature

Neuropsychological rehabilitation remains a fairly new field in most of the world, lagging behind neuropsychological assessment. Yet consequences of brain injury and illness are a major and growing global public health problem (Whiteford et al., 2015), a fact made even more critical given that most low and middle-income countries (LMICs) have few to no neuropsychological rehabilitation services (Watts, 2017). Watts points out that even when services exist, they are often not accessible and/or culturally acceptable to everyone. She goes on to highlight the further complications posed by increasing global migration, globalization, and expanding numbers of refugees. Thus, just as with assessment, there is an issue of whether models and approaches to rehabilitation used in one cultural/linguistic context are appropriate in a completely new context. Furthermore, in

DOI: 10.4324/9781003051497-16

multicultural environments a particular approach to neuropsychological re-habilitation developed to meet the needs of the dominant culture may not be appropriate for all members of that community. Many neuropsychologists will work with patients from cultures and languages that are unfamiliar to them.

Uomoto and Wong (2000) discussed the importance of providing neu-ropsychological services appropriate to the culturally diverse needs of patients in increasingly diverse communities. They argued that cultural sensitivity is perhaps even more important in a rehabilitation setting than in assessment because of the need to not only assess but also to provide support regarding practical issues such as return to work and school or return to driving. Rehabilitation neuropsychologists are likely to be involved in not just diagnosis, but in ongoing interventions, and therefore their need for a broad understanding of cultural factors affecting patients' everyday lives is crucial.

Niemeier et al. (2003) discussed professional codes of conduct and practice relevant to cultural competence for the disciplines of physiatry, rehabilitation psychology/neuropsychology, and nursing in inpatient rehabilitation settings. They identified seven key areas for consideration: (i) continuing education in language and culture; (ii) culturally appropriate assessment tools; (iii) cultural va-lues and beliefs; (iv) recognition of immigration and acculturation experiences; (v) health care and insurance coverage; (vi) attitudes and beliefs about disability; and (vii) past experiences with the medical system. Finally, they provided re-commendations for developing and applying enhanced cultural awareness in clinical rehabilitation practice.

One issue that is relevant to thinking about neuropsychological rehabilitation in different cultures is that resources allocated to neuropsychological rehabilitation vary greatly around the world, and indeed often vary within countries. Even in countries where rehabilitation services are well-resourced, access to services may be much easier in large urban environments than in rural settings. As well as the urban/rural issue, differences in the health care system are also relevant to the accessibility of neuropsychological rehabilitation services. For example, one might expect that in a well-resourced country like the U.S., neuropsychological re-habilitation would be plentiful and available to all. After all, the need is obvious, professional training is generally high quality, professional resources are plentiful, and the country has wealth. And yet, in the U.S., neuropsychologists focus almost exclusively on assessment. Neuropsychological rehabilitation as a term is not widely recognized: graduate programs do not teach rehabilitation, internships and fellowships rarely offer any substantial opportunity to learn rehabilitation, and jobs in neuropsychological rehabilitation are rare. One of the reasons for this is that a billing unit of rehabilitation does not pay as well as a billing unit of assessment. The billing system recognizes cognitive therapy as done by speech and language therapists and by occupational therapists, and many clinical services equate cog-nitive therapy with neuropsychological rehabilitation.

Block et al. (2017) surveyed U.S. neuropsychologists and their involvement in rehabilitation. In this survey, only 173 of 418 neuropsychologists reported engaging in rehabilitation in their practices. They were more likely to provide individual than group therapy, to use technology, and to address both cognitive and psychological issues. Sweet et al. (2002) conducted a practice survey of clinical neuropsychology in the United States and found that cognitive rehabilitation was the least likely clinical activity to be reported, with 81.3% of the sample reporting no rehabilitation activity and for those who did report some rehabilitation activity, the average time per week was 0.82 hours.

Elsewhere in the Americas, other factors influence opportunities for neuropsychological rehabilitation. Arango-Lasprilla et al. (2017) surveyed 808 professionals working in Latin America, mostly in Argentina, Brazil, Colombia, and Mexico. The issues raised by these professionals were consistent with a lag in these countries in the development of neuropsychology services and practice. Barriers included lack of academic training programs and clinical training opportunities, unwillingness to collaborate among professionals, and a lack of culturally relevant and normed tests and other instruments. About half of respondents said they engaged in neuropsychological rehabilitation, and most of these provided rehabilitation services to people with ADHD and/or learning disabilities. Most provided individual therapy and most used personal computers as a technological tool during rehabilitation.

In the recent textbook, *Neuropsychological Rehabilitation: The International Handbook* (Wilson et al., 2017), most of the chapters are written by authors from Europe, Australia, or North America, reflecting the regions of the world where neuropsychological rehabilitation practice and research is most well established. There are also a series of chapters that describe neuropsychological rehabilitation practice in other parts of the world (India, Iran, Russia, Taiwan, Mainland China, Hong Kong, Brazil, Argentina, South Africa, and Botswana). Most of the authors note that neuropsychological rehabilitation practice is less well established in their countries, for a variety of reasons, including a lack of training programs, healthcare systems that do not facilitate access to neuropsychological rehabilitation, a lack of assessment tools for planning rehabilitation, and a lack of translated materials. The accounts from South Africa (Schrieff-Elson and Thomas) and Botswana (Mbakile-Mahlanza and Ponsford) also highlight some specific cultural factors that impact access to rehabilitation. These include cultural beliefs with regard to causes of disability, including a belief that disability arises as a punishment by ancestors for wrongful acts, as an act of God, or by other supernatural causes such as the actions of witches. Disability is stigmatized and there is pressure for families to conceal a family member's difficulties. In the context of these beliefs, treatment may be sought from traditional healers, considered to have the power to tackle the supernatural forces. This approach may limit access to rehabilitation services and evidence-based practices. However, other authors also point out some of the strengths in terms of resources that the community can provide to support rehabilitation. For example, Shah notes that in India large supportive families with a culture of living together and interdependence provide a resource for community-based rehabilitation interventions, with an emerging trend of training family members. This is positive in that it shifts the focus to the recovery of everyday functions that are relevant and of value to the client, though one must be alert for "caregiver burnout, over-protection and over-zealous management by family members" (Wilson et al., p. 503).

My experience of the impact of culture on neuropsychological rehabilitation

I have the probably unique experience of working as a neuropsychologist in three countries with different cultures and levels of wealth, as well as having spent short times as visiting faculty in a large number of other countries, including multiple visits to Russia. I lived and worked in Nicaragua from 1990 to 1992 teaching neuropsychology to practice psychologists, I worked as the Lead Psychologist of the world-renowned Oliver Zangwill Centre for Neuropsychological Rehabilitation in the U.K. from 2010 through 2018, and the rest of my career has been spent in the U.S. working in a variety of settings including medical centers and rehabilitation services.

All four of these countries share the universal problem of large numbers of people with brain injuries and illnesses without enough resources deployed to fully meet the many and varied needs of their populations. Nicaragua, unlike the U.S. and the U.K., is a very impoverished country, second only to Haiti in the western hemisphere. Russia, the U.K. and Nicaragua all have national health systems, and in all three countries, the wealthy have access to private care as well. The Russian health care system was rated in the bottom third of a global ranking of population health in a Lancet report, ranking 119 out of 188 nations. By contrast, the U.K. ranked 5, the U.S. ranked 28, and Nicaragua ranked 100 (Lim et al., 2016).

One of the commonalities among these countries is in the application of neuropsychology knowledge. In all four countries, clinical neuropsychologists are typically engaged primarily in assessment. In many settings in the U.S. and the U.K., the neuropsychologist interviews the patient and provides feedback, while psychology technicians administer and score the tests. Tests are generally part of a set battery with relatively little personalized choice of tests administered. Results may help physicians or other health professionals guide treatment, but patients do not necessarily obtain a great personal direct benefit from the assessment or from the neuropsychologist beyond the feedback session.

The Oliver Zangwill Centre in the U.K.

The Oliver Zangwill Centre (OZC) is part of the National Health Service (NHS) in the U.K. Clients receive intensive neuropsychological rehabilitation that includes a thorough interdisciplinary assessment. The standard program generally involves several weeks of intensive rehabilitation followed by up to several months of community-based rehabilitation and regular follow-ups. The OZC receives funding via the NHS as well as from private sources. There are very few other programs worldwide that offer this intensity of service. The OZC clinicians have a unique opportunity for creativity and innovation of rehabilitation approaches and strategies and for clinical research. For example, Dr. Jessica Fish and a client collaboratively developed an insight-based therapeutic intervention to treat the client's chronic confabulation (Fish & Forrester, 2018). Other examples include examining the therapeutic contribution of a therapy dog to neuropsychological rehabilitation (Winegardner et al., 2015) and training clients to take other people's perspectives as a novel approach to anger management (Winegardner et al., 2016).

The evolution of neuropsychological rehabilitation in Russia

The history of neuropsychological rehabilitation in Russia reflects the strong influence of Soviet-era political and cultural institutions. Monro and Kamaeva (2017) write that there was a strong focus on pride in Russian education and achievements. It has been argued that historically in the Soviet Union persons with physical and mental disabilities were "stigmatized, hidden from the public, and thus made seemingly invisible" (Phillips, 2009). Stigma in relation to disability is not of course unique to the Soviet Union! The ideas, methods, and tools developed by Alexander Luria have been central to neuropsychology assessment and rehabilitation practice in Russia for many decades. Luria's ideas have of course also been influential in many parts of the world, including Latin America where Luria's works were translated into Spanish before the work of European and North American authors and continue to be used in practice (see Solovieva & Quintanar, 2018 for a discussion of the application of Luria's approach to assessment and rehabilitation in a Latin-American context).

In relation to rehabilitation, the main focus in Russia has been on using neuropsychological assessment to build a profile of the patient's cognitive functioning and then selecting a set of impairment-specific cognitive training exercises, with outcome measured in terms of improved performance on cognitive tests. This approach, focused on restoration of impaired cognitive processes, is less favored, at least as a primary approach, in some other countries, where there is more of a focus on teaching strategies to manage (compensate for) cognitive impairment that can be applied to everyday functional situations.

In recent years, colleagues in Russia have been working with colleagues from other countries, including the U.K., to expand neuropsychological rehabilitation knowledge and skills training. Dr. Pauline Monro, Dr. Olga Kamaeva, and other colleagues in St Petersburg and Moscow have played an important part in this work. Dr. Kamaeva is a doctor and physiotherapist in St Petersburg. Dr. Monro is a British neurologist who has fully immersed herself in Russian language, life and culture – she has worked with Russian colleagues to understand rehabilitation practices in Russia and has a passion to help colleagues to consider whether approaches to neuropsychological rehabilitation used in other contexts would be relevant to the Russian context. She has facilitated many exchange visits between Russia and the U.K. and training programs. Her deep connection to Russian culture and colleagues allowed her to work very closely with Russian professionals in planning each new training event to ensure that it was tailored to meet their current needs. Importantly too Dr. Monro also prepared briefings for U.K. colleagues to explain cultural nuances. In relation to "cross-cultural" neuropsychology, this example highlights the importance of not just thinking about the cultural context of patients who are recipients of rehabilitation interventions but also thinking about the cultural context of professionals who are sharing knowledge with each other. This will improve knowledge transfer and guard against neuropsychological imperialism.

I had the good fortune of making five trips to Russia starting in 2015. Each visit garnered more support and interest, and activities grew from involving just an individual lecture to multi-day interactive workshops. Together with my colleague Dr. Jessica Fish, we taught the Oliver Zangwill Centre model and principles of holistic neuropsychological rehabilitation, being sure to emphasize the connection with the ideals and vision of Luria, who is so well respected in Russia and internationally as the founder of clinical neuropsychology. When we attended the Tenth International Congress on Neuro-Rehabilitation in 2018, in Moscow, we were delighted that two Russian colleagues (one psychologist and one neurologist) who had attended our sessions presented cases in which they incorporated the Oliver Zangwill formulation model in their own work, indicating that the model was relevant to the Russian rehabilitation context.

Neuropsychology in post-revolutionary Nicaragua

Nicaragua in the 1980s was a revolutionary country where health care was considered a right and where resources were directed to the poorest in the country. Health posts sprang up across the country and public health played a major role in accomplishments such as the eradication of polio. The field of neuropsychology has its roots in those days with the formation of the Association for Neuropsychology in Nicaragua as the culmination of a seven-week training course offered by the national psychology association in 1988 and taught by two North American neuropsychologists (Dr. Tedd Judd and myself). I followed up by spending over two years from 1990–1992 teaching neuropsychology to a group of practicing psychologists who wanted to gain training in neuropsychology. I brought my North American expectations and tests, we spent time translating and norming them there, and I co-wrote a manual called A Practical Manual of Neuropsychology (Winegardner & Judd, 1992) for use by community health workers.

The hardships I encountered were far worse than I had expected and led me to a new way of thinking about the most effective role of neuropsychology in under-resourced countries. I learned about these hardships in part through supervising two students who carried out a single case study of cognitive rehabilitation at the National Rehabilitation Hospital "Aldo Chavarria". I found the following ten challenges that were true in 1990s Nicaragua and likely remain true in many other parts of the world today:

1. **Inadequate training for psychologists**

   The terminal degree for a licensed professional psychologist in Nicaragua is a bachelor's degree, which requires five years of post-secondary study. This makes it challenging for psychologists to develop specialist knowledge and skills in neuropsychology and neuropsychological rehabilitation to feel confident in assessing, treating, and running research projects with people who have complex neurological conditions. In recent years, a PhD program in Clinical Psychology has been offered at the Central University of Nicaragua, an international school of psychology.

2. **Lack of adequate literature in the local language**

   In the early 1990s, appropriate reference materials, including books and journals, were scarce in Spanish, so that reviewing the literature for information on which to base the students' study was quite difficult. Learning a second language is a luxury in a poor country, and few professionals in Nicaragua speak English. Without an English-speaking neuropsychologist as a consultant, the literature review would have been impossible at that time. There is now much more available neuropsychology literature in Spanish, with translations of Russian works (particularly Luria's writing) as well as North American and European literature, and access to online resources.

3. **Limited availability of human, financial, and material resources**

   There were literally no resources available at that time, whether human, financial, or material, in the hospital, the university, or the country to assist with such projects. The national rehabilitation hospital, which served as the base for the study, had only three telephones which often either did not work properly or went unanswered. The university psychology department had no working telephone. No photocopying was available in the hospital, and copying elsewhere was extremely expensive. There was very little secretarial help. Of course, there were no computers or even many typewriters in the hospital. No funds existed to support the research of this kind, and therefore the researchers had to pay all of the costs involved. With salaries for psychologists at $100 to $120 a month, even a few dollars in costs made research prohibitive.

   Psychologists in Nicaragua did not have access to updated and respected testing materials, and had not had training that would enable them to evaluate the usefulness of existing neuropsychological tests to the local context. For example, the neuropsychologists at the national rehabilitation hospital were gifted the Luria-Nebraska Test Battery, but without relevant norms, this seemingly valuable and impressive-looking gift was not useable.

4. **Inadequate medical information on patients**

   Medical records were often unavailable or inadequate. The only neuro-diagnostic technique available at that time was the EEG. Diagnoses were made clinically without access to more sophisticated testing. Patients often entered the rehabilitation hospital with nothing more than a hand-written chart note from the acute hospital. It was not possible to request medical records because of the poor telephone and mail systems, the lack of photo-copying, and the costs of complying with such requests.

5. **Inadequate conditions for testing and treatment**

   Ideal conditions for treatment (or even minimal conditions, by wealthy countries' standards) did not exist. The offices of the psychologists were thoroughfares through which passed dozens of people, often entering without knocking and interrupting whatever was going on. Because of the lack of privacy and transportation problems, patients were treated in their homes, which usually consisted of one or two rooms shared by many people, often including several children. Quiet and privacy are simply luxuries unavailable in poor countries.

6. **Poor security of materials**

   Theft was rampant in all sectors, and good locks and security were not af-fordable. Some of our testing materials were stolen from one psychologist's office. Since we could not afford more than one set of anything other than paper, we were left without the pictures for our naming test and the timer for a memory test.

7. **Lack of access to transportation**

   No one except the wealthy had personal transport, which meant that ev-eryone, including the psychologists as well as the patients, had to rely on a very overworked and sometimes unreliable bus system for transportation. Moreover, many patients could not afford to come to the hospital for treatment because the bus fare, the equivalent of U.S. 10 cents, was too high.

8. **Economic and personal instability of researchers**

   The two psychology students working on the study also had many personal hardships which made it difficult for them to participate as fully as necessary. Both were working parents as well as full-time students and had serious economic problems. One of them, who ran a stall in the local market, was robbed of all her wares and was then unable to come to our weekly meetings because she had to guard her stall.

9. **Instability of the health system**

   During the course of the study, the hospital, along with the entire national health system, went on strike for six weeks in protest of the meager salaries and lack of medical supplies in the hospitals and clinics, so the study had to be suspended for that time. Later, the patients themselves went on strike briefly, so the hospital was closed and the study was again interrupted.

10. **General economic and political problems**

    The general poverty in the country affected the study in a variety of ways. For example, one patient had to be eliminated because although he met the criteria for the study, he needed eyeglasses in order to participate and could not afford them.

    This experience led me to believe that the prevailing models of neu-ropsychological assessment and rehabilitation in wealthier countries may not fit the realities of low/middle-income countries and that we need to consider the issue of appropriate resource allocation. We should challenge whether it makes sense to continue trying to conduct lengthy and labor-intensive individual cognitive rehabilitation when resources are so scarce and conditions so difficult. The health needs of people with brain injuries are tremendous, and resources are scarce.

An alternative approach might be to direct resources toward prevention and education at the community level. A neuropsychologist, together with any other appropriate professional such as a therapist, nurse, or doctor, could use a patient as an example and teach that person's family, local health workers if available, and community about the patient's disease. For example, the psychologist might evaluate a stroke patient, then meet with that person's family, neighbors, church, and workmates to explain the person's symptoms and problems, suggest management strategies, and teach them about the causes and prevention of stroke. This kind of public health education at the community level may ultimately have much more far-reaching consequences than individual cognitive rehabilitation.

Lead poisoning: an example of community-based rehabilitation

Nicaragua provides an example of the use of neuropsychological test results combined with community involvement and education in reducing lead poisoning in a *barrio* (a very poor neighborhood without paved streets or plumbing) in Managua (Morales Bonilla & Mauss, 1998). In 1991, the *barrio* residents contacted the Ministry of Health with concerns about a car battery factory that belched clouds of black smoke every night from burning empty battery cases. Several cases of severe lead poisoning among the 200 workers had already been reported to the Ministry of Health.

A project was designed with the following aims: (i) to show the magnitude of lead poisoning in the *barrio*; (ii) to educate the local residents and health officials; (iii) to develop a treatment protocol for the affected individuals; and (iv) to minimize overall lead pollution in urban Managua. The project proposal explicitly included social aims including protection of children, education aimed at prevention, and the right of local residents to use the results "according to their own interests". It was one element of a social movement generated to create healthy communities and served as a "guide project" so the experience could be shared with other Managua *barrios*. This project demonstrated how the involvement of local communities in designing the research and implementing changes *ipso facto* meant that local cultural and behavioral mores were integral in every step and ensured local cooperation.

The hypotheses were that IQ and behavior relevant to learning would be impaired in children exposed to lead compared to unexposed children. Before initiating the study, a series of meetings were held in which the *barrio* residents had a voice in determining the nature and goals of the study. The researchers, Morales and

Mauss, also agreed to educate the community about the problem through a series of workshops. The community hoped that the data produced by the study would help convince the health authorities of the seriousness of the problem and persuade them to provide adequate medical attention and take preventive measures. Unsurprisingly, the study found that children exposed to lead had lower IQs and more problem behaviors than unexposed children (Morales Bonilla & Mauss, 1998).

Next, 12 workshops were planned for *barrio* residents, in coordination with the local health post, health brigades and the Community Movement, based on the premise that people with raised awareness are more willing to comply with prevention measures. The workshops provided information about the importance of monitoring lead levels and a description of the effects of lead poisoning. As a result of the community education component, the residents went to the local television station and raised enough public awareness that the battery factory was forced to shut down.

Training of community health workers

Given the shortage of psychologists and the vast mental health needs in LMIC countries, it seems short-sighted to just educate a few psychologists to deal with neuropsychological problems. Many LMIC's have community health workers who treat the basic health needs of the community. These workers usually have limited formal education but are trained in basic health care. Sharing basic knowledge about the recognition and management of brain injury with these community health workers would assure access to at least a minimum level of accurate diagnosis and basic treatment for the vast majority of brain-injured people who currently receive no care whatsoever. Such training can be done through the presentation of the basics of brain injury at a suitable level. Perhaps more important, these workers often have a preventive and public health focus that will have a greater impact than simply treating individual patients.

Psychologists in LMIC countries need more basic training and resources that are relevant to their contexts. Two resources from the same team are: Working with Brain Injury: A Primer for Psychologists Working in Under-Resourced Settings (Coetzer & Balchin, 2014); and Addressing Brain Injury in Under-Resourced Settings: A Practical Guide to Community-Centered Approaches (Balchin et al., 2018). These books are directly aimed at providing a practical, hands-on guide to developing essential competencies and skills in understanding and managing the consequences of brain injury for use at any level of experience in any context, especially those that are under-resourced. They provide basic knowledge of neuropsychology and practical guidance on recognizing and engaging with a wide range of cognitive and emotional/mood problems.

Developing cultural competency

The core theme of this chapter is that it is just as important to consider cultural issues in neuropsychological rehabilitation as it is in neuropsychological assessment. It is imperative that neuropsychologists develop cultural competency in order to provide full, relevant, and acceptable services to all in need of them. De Pereira et al. (2017) acknowledge that very few studies of cultural competency in neuropsychological rehabilitation exist. They draw on resources from other fields including neuropsychological assessment and psychological therapies to set out recommendations for creating services that are adequately attentive to cultural differences. The authors recommend the following in the context of holistic neuropsychological rehabilitation programs:

1. Establish clear and consistent modes of communication within the inter-disciplinary team and between the clients and team.
    This principle applies to all neuropsychological rehabilitation programs, but it highlights the critical importance of working closely with the client and their family to agree on rehabilitation objectives and avoid misunderstanding based on lack of awareness of the culture of the client.
2. Select assessment procedures and instruments that are appropriate to the client's culture and that provide information from a variety of contexts.
    It is essential that the process of assessment, and particularly the interpretation of test results, takes into account the client's cultural context.
3. Incorporate cultural roles and beliefs into the case formulation, goal setting, and rehabilitation interventions.
    Interventions should take account of, and be set in, the cultural context of traditions, social roles, expectations, and beliefs of the individual.
4. Recognize that cultural identity can give a sense of belonging to a group, can provide social support, and should be included when planning community integration.
    Awareness of an individual's cultural context means it may be possible to draw upon appropriate support for the community integration phase of a re-habilitation process.
5. Consider group interactions during intervention planning and delivery.
    An individual's cultural background should be considered in planning a re-habilitation program (setting goals etc.) but also by facilitators during group sessions, and in reflective practice to increase cultural competency.

Conclusions

I have provided a variety of perspectives and reflections on the interface of culture and neuropsychological rehabilitation across four diverse countries. There is no one right approach to implementing neuropsychological rehabilitation practices in any given country given the vast differences in wealth, social and political systems, health systems, and cultures. All countries have in common a large number of people with brain injuries and illnesses who have little access to appropriate, re-levant, meaningful, and helpful neuropsychological rehabilitation for many varied reasons. Individuals who want to create neuropsychological rehabilitation services in a new context are advised to acquire cultural competency in the new location, consult with service users and providers to assure local needs and wishes are met, and advocate for service support at higher levels of institutions and governments. Equally, cultural competency is required when working in multi-cultural contexts, which is increasingly the case in most parts of the world. Solutions will require policy changes at the top levels of government toward valuing this key aspect of health care enough to fund and support it adequately. International collaboration and cooperation will foster the kinds of openness to new ideas and cross-fertilization of experiences that will promote best practices. Finally, adherence to currently established recommendations for cultural competence will assure high standards are met in all contexts and for all people.

References

Arango-Lasprilla, J.C., Stevens, L., Morlett Paredes, A., Ardila, A., & Rivera, D. (2017). Profession of neuropsychology in Latin America. *Applied Neuropsychology: Adult, 24*(4), 318–330.

Balchin, R., Coetzer, R., Salas, C., & Webster, J. (2018) *Addressing brain injury in under-resourced settings: A practical guide to community-centred approaches.* London: Routledge.

Block, C., Santos, O.A., Flores-Medina, Y., Rivera Camacho, D.F., & Arango-Lasprilla, J.C. (2017). Neuropsychology and rehabilitation services in the United States: Brief report from a survey of clinical neuropsychologists. *Archives of clinical neuropsychology, 32*(3), 369–374.

Coetzer, R., & Balchin, R. (2014). *Working with brain injury: A primer for psychologists working in under-resourced settings.* London: Psychology Press.

De Pereira, A.P., Fish, J., Malley, D., & Bateman, A. (2017). The importance of culture in holistic neuropsychological rehabilitation: Suggestions for improving cultural competence. In B.A. Wilson, J. Winegardner, C.M. van Heugten, & T. Ownsworth (Eds.), *Neuropsychological Rehabilitation: The International Handbook.* London: Routledge.

Fish, J. & Forrester, J. (2018). Developing awareness of confabulation through psychological formulation: A case report and first-person perspective, *Neuropsychological Rehabilitation, 28*(2), 277–292.

Lim, S. S., Allen, K., Bhutta, Z. A., Dandona, L., Forouzanfar, M. H., Fullman, N. , ... & Kinfu, Y. (2016). Measuring the health-related Sustainable Development Goals in 188 countries: A baseline analysis from the Global Burden of Disease Study 2015. *The Lancet, 388*(10053), 1813–1850.

McLellan, D.L. (1991). Functional recovery and the principles of disability medicine. In M. Swash & J. Oxbury (Eds.), *Clinical neurology.* Edinburgh: Churchill Livingstone.

Monro, P. and Kamaeva, O. (2017) Neuropsychological rehabilitation in Russia. In B.A. Wilson, J. Winegardner, C.M. van Heugten, & T. Ownsworth (Eds.), *Neuropsychological rehabilitation: The international handbook.* Routledge.

Morales Bonilla, C., & Mauss, E.A. (1998). A community-initiated study of blood lead levels of Nicaraguan children living near a battery factory. *American Journal of Public Health, 88*(12), 1843–1845.

Niemeier, J.P., Burnett, D.M., & Whitaker, D.A. (2003). Cultural competence in the multidisciplinary rehabilitation setting: Are we falling short of meeting needs?. *Archives of Physical Medicine and Rehabilitation, 84*(8), 1240–1245.

Phillips, S.D. (2009) "There are no invalids in the USSR!": A missing Soviet chapter in the new disability history. *Disability Studies Quarterly, 29*, 3. DOI: 10.18061/dsq.v29i3.936.

Solovieva, Y. & Quintanar, L. (2018) Luria's syndrome analysis for neuropsychological assessment and rehabilitation. Psychology in Russia: State of the Art, *11*, 2. doi: 10.11 621/pir.2018.0207.

Sweet, J.J., Peck III, E.A., Abramowitz, C., & Etzweiler, S. (2002). National Academy of Neuropsychology/Division 40 of the American Psychological Association practice survey of clinical neuropsychology in the United States, Part I: Practitioner and practice characteristics, professional activities, and time requirements. *The Clinical Neuropsychologist, 16*(2), 109–127.

Uomoto J.M., Wong T.M. (2000) Multicultural Perspectives on the Neuropsychology of Brain Injury Assessment and Rehabilitation. In Fletcher-Janzen E., Strickland T.L., & Reynolds C.R. (Eds.), *Handbook of cross-cultural neuropsychology. Critical issues in neuropsychology*. Boston, MA: Springer.

Watts, A.D. (2017) Neuropsychological rehabilitation: A global overview. In B.A. Wilson, J. Winegardner, C.M. van Heugten, & T. Ownsworth (Eds.), *Neuropsychological Rehabilitation: The International Handbook*. London: Routledge.

Whiteford, H.A., Ferrari, A.J., Degenhardt, L., Feigin, V., & Vos, T. (2015). The global burden of mental, neurological and substance use disorders: An analysis from the Global Burden of Disease Study 2010. *PLoS ONE, 10*(2), e0116820.

Wilson, B.A., Winegardner, J., van Heugten, C.M., & Ownsworth, T. (Eds.). (2017). *Neuropsychological rehabilitation: The international handbook*. Routledge: London.

Winegardner J., Ashworth F., & Wilson B.A. (2015) The benefits of a therapy dog in a holistic rehabilitation programme. *Neuro-Disability and Psychotherapy, 3*, 11–21.

Winegardner, J., & Judd, T. (1992) *Manual Practica de Neuropsicologia* (Unpublished manuscript distributed in Nicaragua).

Winegardner, J., Keohane, C., Prince, L., & Neumann, D. (2016). Perspective training to treat anger problems after brain injury: two case studies. *Neurorehabilitation, 39*(1), 153–162.

14

THE FUTURE OF NEUROPSYCHOLOGY IN A GLOBAL CONTEXT

Jonathan Evans, Aparna Dutt, and Alberto Luis Fernández

Introduction

Now is an interesting and exciting time to work in the field of neuropsychology. Perhaps this has been said throughout the history of neuropsychology, but just now there are some major challenges to address. Some have suggested the future clinical utility and professional viability of the field is under threat (Cory, 2021). But crises can stimulate change and bring opportunities, and we will argue that we have the opportunity to improve our ability to meet the neuropsychological needs of our global community. This chapter will briefly review the challenges we are facing and then discuss some potential solutions.

The challenges

Neuropsychological theory and practice emerged largely in a European/North American/Australasian context. Inevitably this has meant that neuropsychological tools and practices meet the needs of the majority populations in these regions of the world – essentially white, mostly well-educated people. The neuropsychologists leading the work to develop neuropsychological theory and practice have also largely been white – every single one of the Presidents of the International Neuropsychological Society from 1967 to 2021 has been white (37 men and 17 women). In the most recent survey of professional practices, beliefs, and incomes of U.S. neuropsychologists by Sweet et al., (2021), 84.5% of respondents were Caucasian/White, with just 1.5% African American, 4.5% Hispanic/Latino, and 4.7% Asian/Pacific Islander. These figures are clearly discrepant from the Census 2019 figures for the U.S. population, (60.3% non-Hispanic white; 18.5% Hispanic/Latino; 13.4% African American and 5.9% Asian). Cory (2021) presents a thought-provoking discussion in which he "unpacks the invisible knapsack of

DOI: 10.4324/9781003051497-17

white privilege" highlighting the many ways white privilege affords advantages for a non-Hispanic white neuropsychologist. Cory powerfully argues that understanding white privilege is an important (but limited) step toward improving equity of service for all service-users, whatever their ethnic, cultural, socio-economic, or linguistic background.

Globalization impacts many aspects of our daily lives but in relation to neuropsychology, it is relevant in several key areas. First, migration has meant that many countries of the world that were largely mono-cultural are now truly multi-cultural. This is perhaps best exemplified by the U.S.A., where it is projected that by 2044 the U.S. will become a "majority-minority" nation, in which no ethnoracial group will have a majority share of the total population. In response, the American Academy of Clinical Neuropsychology launched its "Relevance 2050" initiative. This program is highlighting that, as a consequence of the changing demographics in the US, by 2050 "60% of the American population will not be testable with our current neuropsychological toolkit of instruments normed primarily on mono-linguistic/mono-cultural English-speaking European American populations" (American Academy of Clinical Neuropsychology, 2021).

In multicultural contexts, another major issue is the neuropsychology workforce which does not reflect the ethnic/cultural/linguistic backgrounds of the people seeking help from neuropsychology services. This often means that assessment processes must be adjusted, including using interpreters, which can compromise the validity of the assessment (see Chapter 11).

In parts of the world where neuropsychology is less well-established, there is a growing need for neuropsychological services (driven by aging populations, increases in incidence/prevalence of stroke, head injury and other conditions that impact cognition) and so psychologists are looking to develop clinical neuropsychology knowledge and skills. Globalization in terms of improved communication also means that neuropsychological knowledge and tools used elsewhere in the world are more accessible. But many of these tools are not appropriate for use in these different cultural and linguistic contexts. Furthermore, some of these countries are also highly diverse in terms of cultures and languages, India being perhaps the best example, with 31 different languages each spoken by at least a million people. In addition to linguistic diversity, there is a large diversity in ethnicity, religion, culture, education and wealth, all of which challenge the generalizability of "national" norms. So, there are countries in the world without an established tradition of neuropsychology research and practice, where there is a need to develop tools, norms and methods of practice, and in a multicultural context. Use of tests and normative data from places that are culturally and linguistically very different from the one in which they are being used increases the risk of misdiagnosis, ineffective formulation, and inappropriate treatment.

So, there are many challenges facing neuropsychology, but what are the solutions? We do not have any simple solutions to address the access, skills, culture, and availability gaps. Improvements in provision of, and access to, reliable culturally competent neuropsychological services have been made, but progress is slow. Important recommendations and suggestions have been outlined in different articles (Ardila, 1992; Wong & Fujii, 2004) and books (Fletcher-Janzen et al., 2000; Fujii, 2017; Nell, 2000) to aid culturally accurate and appropriate neuropsychological assessment. Here we suggest some practical short-term and long-term approaches. Practical steps will naturally vary according to the individual's socio-cultural context and the political-economic context of the country.

Short-term practical approaches to improve neuropsychological assessment precision

Clinical interviews

During clinical interviews, spend additional time on sociocultural factors including family and society roles and expectations. This also helps in developing a rapport with the patient. To be able to support the person or family to participate effectively in the assessment process it is necessary, as Prigatano (2000) emphasized, to enter the phenomenological world of the patient and her/his family. This applies to any assessment, whatever the background of the patient, but in situations where the patient is from a different cultural or linguistic background to the examiner, or where the experience of being evaluated is unfamiliar, it is especially important to work hard at engaging the patient, putting them at their ease, and facilitating communication. Be aware of the potential for stereotype threat (a person feeling at risk of conforming to stereotypes about their social group). Fujii (2017) illustrates this process with specific recommendations for working (as an American clinical neuropsychologist with an Asian heritage) with patients from Native Indian and Alaskan Native cultures, listing 15 different things that can help create an appropriate cultural context for the assessment. These include things that apply to every assessment, such as being professional, respectful, warm and friendly, and being transparent about the purpose of the assessment, but also include culture-specific guidance such as using the community self-reference for tribal names and being aware of norms for eye contact, style of dress, physical touch, personal space and gender roles (Fujii, 2017, p. 55).

In both diagnostic and rehabilitation assessments, a key issue is to understand what has changed for the patient. What does s/he, or the family, notice is different? What can s/he not do that s/he did previously? What roles have the family taken over from the patient and why? Much of the skill of a clinical neuropsychologist is in observing the patient's behavior and interpreting the clinical history in relation to an understanding of cognitive functions and functional neuroanatomy. Some of the most influential approaches to neuropsychological assessment (such as Luria's analysis of errors – see Ardila (1992); and the Boston Process Approach – see Milberg et al., 2009) emphasize the importance of analyzing performance on tests in relation to models of cognitive processes. But even before administering tests, the neuropsychologist should be analyzing information from the clinical history and observation of behavior. As Ardila (1992) notes, "the neuropsychologist has to command a solid background not only in psychological measurement, but also in neurology, neurophysiology, neuroanatomy and general psychology" (p. 37). Many of the challenges faced by neuropsychology currently are related to use of standardized assessment tests, and while we need to address these issues, it is important to remember that a lot of information relevant to formulating a patient's problems is not dependent on formal test data.

While an analysis of history and interview or everyday behavior contributes greatly to neuropsychological formulation, tests are a critical tool in the assessment process. To interpret test performance accurately it is necessary to obtain all relevant socio-demographic information to inform the selection of appropriate norms, and interpretation of performance in relation to norms. Bear in mind factors that may influence performance but are often not reflected in normative data (e.g., normative data may not include education as a factor but education may be an important factor that influences performance which will be particularly relevant for patients with very low levels of education). Gather information that will help to optimize testing and aid interpretation (e.g., the extent to which drawing, reading, and writing skills are used in everyday life; familiarity with stimuli such as animals or item formats like line drawings vs colored pictures during confrontational naming).

When the patient's educational level is low, it helps if the examiner and the patient belong to the same culture and speak the same language. The implication of this is that all efforts should be made to identify a psychologist from the same cultural and linguistic background to the patient. But this is often not possible and it may be better for a culturally competent neuropsychologist from a different background to see the patient if the alternative is a referral to a psychologist without specialist neuropsychological knowledge and skills even if from a similar background to the patient.

Use of interpreters

In multicultural contexts, and in the absence of a multicultural workforce, the use of interpreters to assist during assessments is increasing. This brings a set of challenges, but there is very useful guidance from a variety of sources which is summarized by Fujii et al. (Chapter 11). Effective management of interpreters to aid assessment is part of being a culturally competent neuropsychologist.

Tests and norms selection

Depending on the purpose of evaluation, plan and select the tests and norms appropriate for the patient's socio-cultural context, educational background, knowledge, skills, and experience. This can be the most challenging task in assessment planning given the lack of availability of tests appropriate to particular cultural and linguistic backgrounds. Search beyond test manuals for evidence that a test is valid in the context in which it is planned to use it. Think carefully about the normative data and the extent to which the normative sample is similar to the patient. It is important to carefully evaluate whether a test has been appropriately developed or adapted even if it is an adaptation of a test that is well-established in a different context.

Testing procedures, test interpretation & reporting

It is good practice in any assessment to administer more than one test assessing the same construct or cognitive domain, but this can be particularly important when there may be concerns regarding whether test performance is being affected by cultural/linguistic issues. More than one test of a domain does not guarantee that performance on multiple tests is not affected by cultural/linguistic issues but improves the chances that an accurate assessment of performance can be obtained. A combination of tests and informant-based-functional scales/questionnaires may help.

Routinely assess language proficiency and level of acculturation when working with bilingual/multilingual individuals, particularly in first- and second-generation immigrants in Western countries. Some standardized acculturation measures are available (Judd et al., 2009; see also Chapter 12). This helps to decide whether an interpreter or a referral to another specialist is required.

Think carefully about interpretation if tests are not administered in a standardized way. For example, if multiple practice trials are given to enable a patient to understand test instructions, interpretation of test results against norms may not be valid. But it is also not appropriate to compare the performance of a patient who does not understand the test instructions with that of a normative sample that fully understands what they have been asked to do.

For some conditions and some domains of functioning a behavioral neurology approach to interpreting performance is entirely appropriate. For example, a patient with visual object agnosia cannot identify objects by sight but may be able to identify them by touch - visual object identification is not a continuously distributed cognitive variable and in this case, it is reasonable that the neuropsychologist classifies this symptom as present/absent (though may also have information available on a standardized naming test). However other cognitive functions, such as memory, are continuously distributed in the population and hence it is possible to interpret the performance of a patient against that of the normative sample. Good normative data includes people with all of the sociodemographic characteristics that are known to affect test performance. Accurate interpretation needs normative data to be available in a form that allows the patient's performance to be compared with a group of people who are similar in all of the factors that are known to influence performance. This could be via stratified norms (e.g., stratified by age, sex, education, etc) or via regression-based norms that look at the level of discrepancy from expected performance based on demographic characteristics. It is important to document the assumptions that have been made in relation to interpreting test results, particularly when the normative sample does not match the patient's characteristics well.

There are many factors that can affect test performance in addition to the specific cognitive function that a test is designed to assess. It is important to consider background history, cultural and contextual factors, behavioral observations, test-taking skills, effort level, test behavior, qualitative performance, and medical history. Check how well the test findings (especially low scores) fit well with clinical observations and if inconsistent don't assume that your clinical judgment is poor – it may be the tests that are poor.

It is important to note any issues related to testing procedures (e.g., deviation in test administration, use of interpreters, whether the patient was allowed to use any specific strategies to perform optimally, norms used) and all limitations should be clearly outlined in the report.

Referral, peer consultation, and collaboration

Clinicians may naturally wish to refer to, or consult, other experts who may be more familiar with the culture and language of the patient in question. However, given the shortage of qualified culturally and linguistically diverse neuropsychologists, this can be problematic.

Virtual collaborative clinical practice can help to provide more equitable neuropsychological services both within and across countries.

In this section, we have considered suggested some short-term solutions that may increase the likelihood of doing a culturally competent neuropsychological assessment. But there are longer-term issues that need longer-term solutions and these are considered next.

Long-term practical approaches to reduce problems of access to neuropsychological services and improve neuropsychological assessment precision

Increasing awareness and acceptability of neuropsychological services in the general population

In contexts where neuropsychology services are not well established or where some groups within a community are underserved, ways to raise awareness of neuropsychological services and also improve their acceptability must be determined. Engaging and collaborating with patient advocacy organizations may help. In LMICs, training people in the community for early identification may help to raise awareness and eventually acceptability of neuropsychological services.

Accessibility of neuropsychological services

In many Western countries where neuropsychology is well-established, there is a need to train and recruit a more diverse workforce to address the needs of a culturally diverse population (Echemendia et al., 1997). This work needs to start at the level of schools and undergraduate levels at the University to encourage people from diverse cultural backgrounds to consider a career in clinical/neuropsychology. In a small profession there may never be sufficient diversity for all patients to be seen by someone from the same background unless novel approaches to service delivery are considered (e.g., patients being seen by a psychologist from out of their local area, perhaps via a consulting program where psychologists from under-served communities provide consultation, or direct service, across multiple areas in a country).

Developments and innovations, such as teleneuropsychology/app-based neuropsychological assessment, may help access remote populations (see Hammers et al. (2020)). However, the extent to which such technologies are viable and useful for culturally, linguistically and educationally diverse populations across the world needs to be explored.

Access to neuropsychological tests

There is a lack of reliable and valid neuropsychological tests with appropriate normative data in many parts of the world. Many of the chapters in this book have highlighted that culture (in the broadest use of the term) influences performance on cognitive tests. This might be in relatively subtle ways, such as the impact of belonging to an individualist/collectivist culture on attention and memory test performance, or in very obvious ways, such as a naming test that includes items that are not familiar to the population in the new context. A clear implication is that it is inappropriate to take any test that is used in one context and simply translate the instructions and materials and use it in a new context, without examining whether the test is reliable and valid in the new context. In most cases, some form of adaptation of materials or procedures is required.

The International Test Commission guidelines for translating and adapting tests (ITC, 2017) is a useful starting point for thinking about issues of test adaptation. These guidelines document the processes involved in adapting tests for use in a new context. But even before considering adapting an existing test, it is important to think carefully about what you want a test to measure. In relation to cognitive functions, how does an impairment in this function manifest itself in the new context? For example, if you want to develop a test to detect memory impairment associated with dementia, what memory difficulties seem to be present in people with dementia in your context? Could a test be developed "from scratch"? Initial studies need to understand the nature of memory impairments to identify what type of test might be useful. If a test can be developed to assess the construct of interest it may be better to do that than adapt an existing test. However, many tests in countries with well-established neuropsychology have undergone decades of research and clinical usage, so it will often make sense to adapt a test. This may require discussion with test publishers and some legal hoops to jump through. But it is worth pursuing as most test publishers are committed to serving the global community, especially if there is a new market. It is important for someone who is competent in the language of the original test and the language used in the new context, and understands the construct being measured to consider how much of the test can be simply translated and how much needs to be adapted. At every stage, it is important to consider whether it is appropriate to continue with the adaptation. The materials and instructions should be translated but it is important that the translation process focuses on translating for neuropsychological meaning rather than literal meaning. If material needs to be adapted then the rationale for the selection of new items/materials must be clear. Extensive research in developing new materials is often required. Back translation may be relevant for task instructions, but will often not be relevant for adapted materials. Pilot studies are required to determine whether materials are understandable to examiners and patients and whether the test appears to be measuring what it is intended to measure. The pilot studies provide the opportunity to stop the adaptation process at an early stage if the test is not working. Next, a formal evaluation of reliability and validity is required. Keep in mind that validity refers to the extent to which the test is useful for the purpose for which it is intended.

Finally, normative data can be collected. Normative samples should reflect all of the key characteristics that are known to, or considered likely to, affect test performance. In many of the world's major neuropsychological tests, it has traditionally been age and sex that have been the focus of normative data stratification. Education is also important. In most high-income countries, the impact of years of education may be lessening (with most young people being required to stay in education through to 16 or 18 years) but the quality of education is critical (see Chapter 4). In LMIC's, although minimum levels of education are increasing there remains a large proportion of the population with very little formal education and low levels of literacy. Developing literacy-based norms in addition to age, education, and sex is helpful particularly in countries with high illiteracy (Dotson et al., 2009; Youn et al., 2011). Regression-based norms are becoming increasingly popular, especially with access to computerized scoring systems, as a variety of demographic factors can be considered at the same time. But more work is needed to understand the determinants of brain health and cognitive performance and ensure that appropriate normative data sets are used that neither under-or over-estimate cognitive impairment (Possin et al., 2021).

A "Second Adaptation" or "Third norm" version is encouraged for tests that have been adapted and normed in a new context but need to be adapted again to assess migrant groups (e.g., Addenbrooke's Cognitive Examination III − Bengali version adapted from the original English version for Indian Bengali's but requires further adaptation for British Bengalis).

Universal tests − the holy grail of neuropsychology?

One solution to the issue of the impact of culture on cognitive performance is to develop culture-free, or universal tests (Fernandez & Abe, 2018; Fernandez & Marcopulos, 2019). There have been several attempts to develop tests that work well across cultures, with minimal adaptation. However, these are largely confined to relatively brief screening tools (e.g., RUDAS; Storey et al., 2004) and no tool is completely free from the impact of sociodemographic and cultural factors. For example, Nielsen et al. (2018) examined the performance of people from a western European majority population as well as from five different minority populations. They found significant performance differences between the ethnic groups and the majority population sample, though the differences were mainly explained by differences in education with ethnicity and acculturation explaining only a small amount of variance. It appears then that there is scope for reducing cultural bias, but it is not possible to remove bias completely, and education remains a critical factor in test performance.

Training

In countries with well-established neuropsychology services, it is important to integrate cultural competency training into clinical neuropsychology training programs, including clinical case presentations involving minority patients, and supervision by clinicians with multicultural competence (Echemendia et al., 1997).

In the absence of academic clinical training programs in neuropsychology in many LMICs and non-Western high-income countries, flexible training pathways that are pragmatic in the country's context may be needed. This may include (a) structured and extensive supervised clinical training/internship programs to supplement academic qualifications (b) attending online clinical mentorship provided by national or international experts and (c) attending national/international continuous education knowledge and skill-based training workshops. Training should address educational, cultural, and linguistic issues that are appropriate to the population in question.

Collaborations and research

To facilitate culturally appropriate clinical practice or fairness in testing, more cross-cultural neuropsychology research, including collaborations with linguists and cognitive neuroscientists is vital. Integrating advances in cultural neuroscience into clinical neuropsychology could potentially drive the development of novel assessment paradigms and improve precision in cross-cultural neuropsychological assessment. International cooperation is vital for establishing competency and capacity in clinical neuropsychology and research in countries where neuropsychology is not well developed. Cross-cultural clinical experiences and insights of experts from LMICs are critical in international collaborative research projects (Noroozian et al., 2014). Data transparency related to testing translation/adaptation/development methods should be made mandatory for academic publications.

Moreover, it is necessary to more accurately determine what specific aspects of culture are affecting cognitive functions and neuropsychological performance.

Where the influence of culture is observed, it is important to determine how this influence is exerted. For example, do different languages influence the expression of executive functions? If the answer is yes, is this influence equally expressed in all languages, does it affect some families of languages, or is this expression present in some specific languages but not in others?

National neuropsychological societies

Developing local neuropsychological societies in non-western nations is vital for understanding regional needs and challenges, and achieving many of the afore-mentioned goals, including identifying national experts, establishing training pathways and credentialing systems, organizing conferences and continuing education events, and setting standards for clinical practice and research. This will eventually propel the development of neuropsychology as an independent discipline, and facilitate its integration into the country's healthcare system.

Conclusion

Neuropsychology must adapt to meet the needs of a globalized world. We need a more diverse workforce and culturally competent practitioners. Cultural neuroscience is critical to understanding brain-behavior relationships and what factors we must address when developing, adapting, and interpreting tests. In places without tests, development from scratch is ideal (and will stimulate creative approaches and more precise assessment), but where existing tests are adapted to a new context then robust adaptation methods must be applied. Neuropsychological rehabilitation interventions must also be relevant to the cultural context of the individual – one size will never fit all.

There is much to do, and these ideas are just a start, but there is clearly a willingness and enthusiasm amongst neuropsychologists to prioritize this work to improve our ability to serve our global community better.

References

American Academy of Clinical Neuropsychology. (2021). *Relevance 2050*. https://theaacn. org/relevance-2050/ Accessed 5th August 2021.

Ardila, A. (1992). Luria approach to neuropsychological assessment. *International Journal of Neuroscience, 66*(1–2), 35–43. doi:10.3109/00207459208999787.

Cory, J.M. (2021). White privilege in neuropsychology: An 'invisible knapsack' in need of unpacking? *Clinical Neuropsychologist, 35*(2), 206–218. 10.1080/13854046.2020.1801845.

Dotson, V.M., Kitner-Triolo, M.H., Evans, M.K., & Zonderman, A.B. (2009). Effects of race and socioeconomic status on the relative influence of education and literacy on cognitive functioning. *Journal of the International Neuropsychological Society, 15*, 580–589. doi:10.1017/S1355617709090821.

Echemendia, R.J., Harris, J.G., Congett, S.M., Diaz, M.L., & Puente, A.E. (1997).

Neuropsychological training and practices with Hispanics: A national survey. *Clinical Neuropsychologist, 11*(3), 229–243. 10.1080/13854049708400451.

Fernández, A. L., & Abe, J. (2018). Bias in cross-cultural neuropsychological testing: Problems and possible solutions. *Culture & Brain, 6*, 1–35.

Fernandez, A. L. , & Marcopulos, B. A., (2019). Cross-cultural tests in neuropsychology: A review of recent studies and a modest proposal. InSandra, Koffler, E. Mark, Mahone, Bernice, Marcopulos, Douglas, Johnson-Greene, & Glenn, Smith (Eds.). *Neuropsychology: a review of science and practice III* (pp.93–128). New York: Oxford University Press Series.

Fletcher-Janzen, E., Strickland, T.L., & Reynolds, C.R. (Eds.). (2000). *Critical Issues in Neuropsychology. Handbook of Cross-Cultural Neuropsychology.* Kluwer Academic Publishers. 10.1007/978-1-4615-4219-3.

Fujii, D. (2017). *Conducting a culturally informed neuropsychological evaluation.* Washington, DC: American Psychological Association.

Hammers, D.B., Stolywk, R., Harder, L., & Cullum, C.M. (2020). A survey of international clinical teleneuropsychology service provision prior to and in the context of COVID-19. *Clinical Neuropsychologist, 34*(7–8), 1267–1283. doi:10.1080/13854046.2020.1810323.

International Test Commission. (2017). *The ITC guidelines for translating and adapting tests* (2nd ed.). [www.InTestCom.org].

Judd, T., Capetillo, D., Carrion-Baralt, J., Marmol, L.M., Miguel-Montes, L.S., Navarrete, M.G., ... Planning, C. (2009). Professional considerations for improving the neuropsychological evaluation of hispanics: A National Academy of Neuropsychology Education paper. *Archives of Clinical Neuropsychology, 24*(2), 127–135. doi:10.1093/arclin/acp016.

Milberg, W.P., Hebben, N., & Kaplan, E. (2009). The Boston Process Approach to neuropsychological assessment. In I. Grant & K.M. Adams (Eds.), *Neuropsychological assessment of neuropsychiatric and neuromedical disorders* (pp. 42–65). Oxford University Press.

Nielsen, T.R., Segers, K., Vanderaspoilden, V., Bekkhus-Wetterberg, P., Minthon, L., Pissiota, A., ... Waldemar, G. (2018). Performance of middle-aged and elderly European minority and majority populations on a Cross-Cultural Neuropsychological Test Battery (CNTB). *Clinical Neuropsychologist, 32*(8), 1411–1430. doi:10.1080/13854046.2018.1430256.

Nell, V. (2000). *Cross-cultural neuropsychological assessment: theory and practice.* Mahwah, NJ: Lawrence Erlbaum Associates, Inc.

Noroozian, M., Shakiba, A., & Iran-Nejad, S. (2014). The impact of illiteracy on the assessment of cognition and dementia: A critical issue in the developing countries. *International Psychogeriatrics, 26*(12), 2051–2060. 10.1017/S1041610214001707.

Possin, K.L., Tsoy, E., & Windon, C.C. (2021). Perils of race-based norms in cognitive testing the case of former NFL players. *JAMA Neurology, 78*(4), 377–378. doi:10.1001/jamaneurol.2020.4763.

Prigatano, G.P. (2000). Neuropsychology, the patient's experience, and the political forces within our field. *Archives of Clinical Neuropsychology, 15*(1), 71–82. doi:10.1016/s0887-6177(99)00021-9.

Storey J., Rowland J., Basic D., Conforti D., & Dickson H. [2004] The rowland universal dementia assessment scale (RUDAS): A multicultural cognitive assessment scale. *International Psychogeriatrics, 16*(1) 13–31

Sweet, J.J., Klipfel, K.M., Nelson, N.W., & Moberg, P.J. (2021). Professional practices, beliefs, and incomes of U.S. neuropsychologists: The AACN, NAN, SCN 2020 practice and "salary survey", *The Clinical Neuropsychologist*, *35*(1), 7–80, DOI: 10.1 080/13854046.2020.1849803.

Wong, T. M., & Fujii, D. E. (2004). Neuropsychological assessment of Asian Americans: Demographic factors, cultural diversity, and practical guidelines. *Applied Neuropsychology*, *11*(1), 23–36. 10.1207/s15324826an1101_4.

Youn, J., Siksou, M., Mackin, R.S., Choi, J., Chey, J., & Lee, J. (2011). Differentiating illiteracy from Alzheimer's disease by using neuropsychological assessments. *International Psychogeriatrics*, *23*(10), 1560–1568. 10.1017/S1041610211001347.

INDEX